Springer Series in Advanced Microelectronics

Volume 39

Series Editors

Dr. Kiyoo Itoh, Kokubunji-shi, Tokyo, Japan
Professor Thomas H. Lee, Stanford, CA, USA
Professor Takayasu Sakurai, Minato-ku, Tokyo, Japan
Professor Willy M. Sansen, Leuven, Belgium
Professor Doris Schmitt-Landsiedel, Munich, Germany

For further volumes:
www.springer.com/series/4076

The Springer Series in Advanced Microelectronics provides systematic information on all the topics relevant for the design, processing, and manufacturing of microelectronic devices. The books, each prepared by leading researchers or engineers in their fields, cover the basic and advanced aspects of topics such as wafer processing, materials, device design, device technologies, circuit design, VLSI implementation, and subsystem technology. The series forms a bridge between physics and engineering and the volumes will appeal to practicing engineers as well as research scientists.

Richard Gaggl

Delta-Sigma A/D-Converters

Practical Design for Communication Systems

 Springer

Richard Gaggl
Design Center Villach
Infineon Technologies
Villach, Austria

ISSN 1437-0387   Springer Series in Advanced Microelectronics
ISBN 978-3-642-43599-7   ISBN 978-3-642-34543-2 (eBook)
DOI 10.1007/978-3-642-34543-2
Springer Heidelberg New York Dordrecht London

Printed on acid-free paper

Springer is part of Springer Science+Business Media (www.springer.com)

For Tine, Marie and Anna

Preface

Analog and mixed-signal circuits continue to play a major role in modern technologies. Obviously, interfacing with antennas, wires, microphones and image sensors requires analog signal processing such as signal amplification, filtering and conversion from analog to digital. A/D-converters are bridges between the analog real world and the digital world. In practical applications the A/D-converters often become bottlenecks and their design is critical to overall system performance. Furthermore, the A/D-converters show a significant impact on system efficiency (power consumption), product costs (chip area) and robustness. Nevertheless, digital circuits drive the market such as microprocessors and system-on-chip solutions. The technology development has been optimized over the last decades to reduce the power consumption and chip area of digital circuits. Contrarily, the implementation of analog circuits has become more difficult using these modern semiconductor technologies. Smaller transistor sizes come along with a reduced tolerable voltage range resulting in a more challenging design. As a consequence, modern CMOS technologies drive the need for novel architectures to keep pace with improving efficiency of A/D-conversion.

The emphasis of this book is on practical design aspects for broadband analog-to-digital converters for communications. The embedded designs are employed for transceivers in the field of Asymmetric Digital Subscriber Line (ADSL) solutions and Wireless Local-Area Network (WLAN) applications. An area- and power-efficient realization of a converter is mandatory to remain competitive in the market. The right choice for the converter topology and architecture needs to be done very carefully to result in an attractive figure-of-merit (FOM). The book begins with a brief overview of basic concepts about ADSL and WLAN to understand the requirements on Analog-to-Digital (A/D) converters. In both fields of applications it will turn out that a delta-sigma approach is a good choice for fulfilling all requirements. At architecture level, issues on different modulator topologies are being discussed employing the provided technology node. The technology features influence the choice of different architectures in order to improve the circuit implementation. The design issues are pointed out in detail for modern digital Complementary Metal-Oxide-Semiconductor (CMOS) technologies, beginning with 180 nm and going down to 65 nm feature size. Beside practical aspects, challenges on mixed-signal design level are addressed to optimize the converters in terms of consumed chip area, power consumption and design for yield in high volume production. Thus, careful considerations on circuit- and architectural-level are done by introducing a dynamic-biasing technique, a feedforward-approach and a resolution in time instead of amplitude resolution. In 180 nm CMOS, a 85 dB dynamic range multi-bit delta-sigma A/D converter as well as in 130 nm CMOS, a power optimized 14 bit switched-capacitor delta-sigma modulator, are designed for ADSL applications, achieving $1.8 \frac{pJ}{conv}$ and $0.7 \frac{pJ}{conv}$, respectively. In 65 nm CMOS, a design for a 10 bit delta-sigma converter with a signal bandwidth up to 20 MHz is presented for a baseband WLAN solution attaining $0.15 \frac{pJ}{conv}$. A new architecture is introduced by a time-encoding technique replacing a conventional amplitude quantization.

Acknowledgements

The content of the book reflects my work as a mixed-signal design engineer in the former communications group at Infineon Technologies and my work as principal engineer at Lantiq, Villach, Austria. Furthermore, the content was enhanced during my stay at the Technical University of Graz, Austria, for working on my doctoral thesis. This book would not have been possible without the help and support from many people.

There are many people to thank for supporting the work that has been compressed into this book. At this point, I would like to thank the people who directly contributed to the content of this book. First of all, I owe deep gratitude to Prof. Dr. Wolfgang Pribyl for inspiration and the initial encouragement to carry out this book, as well as his supervision and help during my stay at the Technical University of Graz. I am deeply indebted to my co-supervisor Prof. Dr. Willy Sansen from the Katholieke University Leuven whose help, stimulating suggestions on all technical aspects and encouragement helped me in all the time of research for and on the general style of writing the manuscript. Furthermore, I would like to thank Assoc. Prof. Dr. Luis Hernandez, Assoc. Prof. Dr. Susana Paton and Ass. Prof. Dr. Enrique Prefasi, all from the University Madrid, for their stimulating ideas and academic background investigations on time-encoding quantizers.

I would like to thank both companies Infineon Technologies and Lantiq for allowing me to concentrate the results of several years of working experience in this exciting field into a volume for the series of Advanced Microelectronics. Thanks to my colleagues from the communications department for their help with design, layout and measurements on my three test-chips. I want to thank them for all their help, support, interest and valuable hints. Further on, I especially want to thank Dr. Andreas Wiesbauer for his great help with oversampling converters as well as Dr. Dietmar Sträußnigg for his support in control engineering and the needed boost throughout my work.

Lastly, and most importantly, I would like to give my special thanks to my whole family. Especially, I wish to express my gratitude to my wife Tine and my daughters

Marie and Anna whose patient love enabled me to complete this work and for their understanding of my absence during long working weekends. To them I dedicate this book.

Kärnten, Austria   Richard Gaggl

Contents

List of Notations

Acronymus and Abbreviations

ADSL	Asymmetric Digital Subscriber Line
ADSL2+	Asymmetric Digital Subscriber Line 2+
A/D	Analog-to-Digital
ADC	Analog-to-Digital Converter
AFE	Analog Front End
ASIC	Application-Specific Integrated Circuit
BSS	Basic Service Set
CMOS	Complementary Metal-Oxide-Semiconductor
CO	Central Office
conv	conversion
CPE	Customer Premises Equipment
CT	Continuous Time
DAC	Digital-to-Analog Converter
D/A	Digital-to-Analog
DC	Direct Current
DEM	Dynamic Element Matching
DLC	Digital Loop Carrier
DMT	Discrete Multi-Tone
DNL	Differential Non-Linearity
DR	Dynamic Range
DSL	Digital Subscriber Line
DSSS	Direct-Sequence Spread-Spectrum
DT	Discrete Time
ELD	Excess Loop Delay
ENOB	Effective Number-of-Bits
FET	Field-Effect Transistor
FFT	Fast Fourier Transform
FHSS	Frequency-Hopping Spread-Spectrum
FIR	Finite Impulse Response

FOM	Figure Of Merit
GBW	Gain-BandWidth Product
GOX	Gate-Oxide
HDSL	High bit-rate Digital Subscriber Line
IBQN	In-Band Quantization Noise
ICN	Idle Channel Noise
IEEE	Institute of Electrical and Electronics Engineers
INL	Integral Non-Linearity
ISDN	Integrated Service Digital Network
ITU-T	International Telecommunications Union-Standardization Sector
JFET	Junction Field-Effect Transistor
LAN	Local-Area Network
MAC	Media Access Control
MASH	Multi-stAge noise-SHaping
MOSFET	Metal-Oxide-Semiconductor Field-Effect Transistor
MTPR	Missing-Tone Power Ratio
NTF	Noise Transfer Function
OFDM	Orthogonal Frequency-Division Multiplexing
OpAmp	Operational Amplifier
OSI	Open System Interconnection
OSR	Over-Sampling Ratio
OTA	Operational Transconductance Amplifier
PLL	Phase-Locked Loop
POTS	Plain-Old Telephone Service
PSD	Power Spectral Density
PSTN	Public Switched Telephone Network
PWM	Pulse-Width Modulation
RAM	Random-Access Memory
rms	root mean square
ROSR	Ratio-OSR
SAR	Successive Approximation Register ADC
SC	Switched-Capacitor
SFDR	Spurious-Free Dynamic Range
S/H	Sample-and-Hold
SNDR	Signal-to-Noise and Distortion Ratio
SNR	Signal-to-Noise Ratio
SoC	System on Chip
SPICE	Simulation Program with Integrated Circuit Emphasis
SQNR	Signal-to-Quantization-Noise Ratio
STF	Signal Transfer Function
TCP/IP	Transmission Control Protocol/Internet Protocol
TDC	Time-to-Digital Converter
TEM	Time-Encoding Machine
TEQ	Time-Encoding Quantizer

THD	Total Harmonic Distortion
VLSI	Very Large-Scale Integrated Circuits
WLAN	Wireless Local-Area Network

Physical Symbols

C	Capacitance [F]
f	Frequency [Hz]
K	Boltzmann's constant (1.38×10^{-23}) [J/K]
I	Current [A]
P	Power (dissipation) [W]
R	Resistance [Ω]
t	Time [s]
T	Absolute temperature [K]
V	Voltage [V]

Electrical Symbols

α	(Relativ) pulse width [–]
a_i	Polynomial coefficient [–]
A_{max}	Amplitude of a sinusoidal signal [V]
A_0	Finite OTA or OpAmp gain at DC [dB]
B	Number of quantization bits [–]
BW	Bandwidth [Hz]
C	Number of relevant clock edges within one period ($C = 1$ or $C = 2$) [–]
C_{eq}	Equivalent capacitance [F]
C_I	Integrator capacitance [F]
C_L	Load capacitance [F]
C_P	Parasitic capacitance [F]
C_S	Sampling capacitance [F]
$conv$	Conversion [–]
Δ	Quantizer step size [V]
Δ_3	Distortion coefficient [–]
$e(n)$	Discrete-time error signal [V]
$E(z)$	Discrete-time quantization error in z-domain [V]
$E_1(z)$	Discrete-time quantization error in z-domain of first quantizer [V]
$E_2(z)$	Discrete-time quantization error in z-domain of second quantizer [V]
f_B	Analog signal bandwidth [Hz]
f_S	Sampling frequency [Hz]
f_N	Nyquist frequency [Hz]
f_{sig}	Signal frequency [Hz]
G_E	Gain Error of an integrator [dB]
g_{lp}	Gain of a lowpass filter [–]
g_m	Transconductance of a transistor [A/V]
g_o	Output conductance of a transistor [A/V]
$H_D(z)$	Discrete-time transfer function of a digital filter [–]
$H_{OP}(s)$	Continuous-time transfer function of an OpAmp [–]

$H(s)$	Continuous-time (loop-filter) transfer function [–]
$H(z)$	Discrete-time (loop-filter) transfer function [–]
$\Im\{H(j\omega)\}$	Complex part of transfer function $H(j\omega)$ [–]
$h_a(t)$	Continuous time (analog) impulse response [V]
$h_d[n]$	Discrete time (digital) impulse response [V]
$J_p(x)$	Bessel function of the first kind and order p [–]
k	Index number [–]
k_g	Coefficient for gain-scaling [–]
K'	Gain factor of a MOS-device [A/V^2]
K_c	Unity gain frequency of an integrator [Hz]
λ_1	Capacitive feedback factor during sampling phase [–]
λ_2	Capacitive feedback factor during integrating phase [–]
L	Length of a MOS-device [m]
L	Order of loop filter in a delta-sigma modulator [–]
$L^{-1}\{H(s)\}$	Inverse Laplace transformation of $H(s)$ [–]
μ	Frequency of modulating input signal for PWM [Hz]
m	Index number [–]
N	Number of bits of a quantizer [–]
$\phi(f)$	Phase noise density [V^2/Hz]
ϕ_1	First (sampling) phase of switched capacitor circuit [–]
ϕ_{1d}	First (sampling) delayed phase of switched capacitor circuit [–]
ϕ_2	Second (integrating) phase of switched capacitor circuit [–]
ϕ_{2d}	Second (integrating) delayed phase of switched capacitor circuit [–]
$p(t)$	Continuous-time pulse [V]
p_i	Frequency of a pole [Hz]
$\hat{P}$	Amplitude of pulse $p(t)$ [V]
ω	Angular frequency [rad/s]
ω_0	Center frequency [rad/s]
ω_c	Carrier frequency [rad/s]
ω_D	Dominant pole frequency of an OpAmp [rad/s]
ω_{lp}	Pole frequency of a lowpass filter [rad/s]
ω_{osc}	Limit-cycle frequency within an oscillating PWM loop [rad/s]
ω_{sig}	Signal Frequency [rad/s]
$\hat{P}$	Amplitude of a square waveform [V]
P_N	Noise power [W]
Q	Quality factor of a pole or zero [–]
$\Re\{H(j\omega)\}$	Real part of transfer function $H(j\omega)$ [–]
$r_x(0)$	Autocorrelation function of function x at zero [–]
$r_{tj}(0)$	Autocorrelation function of jitter function t_j at zero [–]
R_I	Input resistance [Ω]
R_L	Load resistance [Ω]
Ron	Channel-resistance of a MOS in linear region [Ω]
Rsw	On-resistance of a closed switch [Ω]
ρ_1	Static error for closed loop configuration [–]
ρ_2	Static error for closed loop configuration [–]

s	Laplace-domain variable [rad/s]
s_{LP}	Parasitic pole frequency of a real integrator [rad/s]
$S_A(f)$	Power spectral density of the analog signal [W/Hz]
$S_e(f)$	Power spectral density of the error (quantization) noise [W/Hz]
$S_\phi(f)$	Power spectral density of the phase noise [W/Hz]
$S_q(f)$	Power spectral density of the shaped quantization noise [W/Hz]
$S_{tj}(f)$	Power spectral density of the jitter function [W/Hz]
σ_{abs}	Absolute Jitter [s]
$\sigma_{acc}(m)$	Accumulated jitter [s]
σ_e^2	Variance of uniformly distributed (interpolation) error [V^2]
σ_{lt}	Long-term jitter [s]
σ_{per}	Period Jitter [s]
σ_q^2	Variance of uniformly distributed quantization error [V^2]
SNR_j	Signal-to-noise ratio due to sampling with clock jitter [–]
t_j	Jitter function (time discrete random process) [–]
T	Signal period [s]
T_c	Limit-cycle period [s]
T_{osc}	Limit-cycle period within an oscillating PWM loop [s]
T_s	Sampling period [s]
T_{sig}	Signal period [s]
$T[k]$	Irregularly sampling sequence [–]
$u(t)$	Continuous-time signal [V]
$\hat{U}$	Amplitude sinusoidal signal [V]
v	Variable voltage [V]
v_m	Voltage amplitude [V]
V_D	Node voltage at drain of a MOS device [V]
V_G	Node voltage at gate of a MOS device [V]
V_i	Input (node) voltage [V]
V_o	Output (node) voltage [V]
V_{ref}	Provided reference voltage for A/D conversion [V]
V_S	Node voltage at source of a MOS device [V]
Vth	Threshold voltage of a MOS-device [V]
$w(t)$	Continuous-time signal [V]
W	Width of a MOS-device [m]
$x[m]$	Regularly sampled signal [V]
$x(n)$	Discrete-time (input) signal [V]
$x(t)$	Continuous-time (input) signal [V]
$x(T[k])$	Irregularly sampled signal [V]
$X(z)$	Discrete-time (input) signal in z-domain [V]
$y(n)$	Discrete-time (output) signal [V]
$y(t)$	Continuous-time (output) signal [V]
$Y(z)$	Discrete-time (output) signal in z-domain [V]
$Y_1(z)$	Discrete-time (output) signal in z-domain of first stage [V]
$Y_2(z)$	Discrete-time (output) signal in z-domain of second stage [V]

z	Discrete-time frequency variable in z-domain, $z = e^{sT_S}$ [V]
z_i	Frequency of a zero [rad/s]
$Z\{h_d[n]\}$	z-transformation of $h_d[n]$ [–]
$Z^{-1}\{H(z)\}$	Inverse z-transformation of $H(z)$ [–]

Chapter 1
Introduction

1.1 Motivation

The second half of the 20th century has been a remarkable period for technological innovation, and particularly so for telecommunications. During the first half of the last century, telecommunications was almost entirely in the analog mode. Digital transmission of voice and low-speed data communications were the first ventures into digital transmission along a twisted pair telephone line. The invention of the transistor and the subsequent innovation of integrated circuits quickly led to a revolution in electronics and in communications in general. The rapid development of computer and consumer electronics, traditional telephone subscribers are ready for new services based on digital technologies. Driven by the deregulation in the telecommunications industry and the vast growth rate of internet users, world wide industry has made enormous efforts to deploy and standardize a multitude of various different access technologies ranging from media such as air, twisted-pair copper cables, coax cables and optical fibers. Applications such as video conferencing, fast internet downloads, digital television and tele-working have been made available to the public with wide coverage. With a powerful PC processing lots of information, the capability of a voice band modem becomes a limiting factor. The voice channel is limited by the 3.1 kHz band bottlenecking the achievable data-rate for integrating consumers to the evolving high-speed digital communication network. The drawback of a telephone modem can be overcome with the Digital Subscriber Line (DSL) technology. There were several key innovations such as Integrated Service Digital Network (ISDN) and High bit-rate Digital Subscriber Line (HDSL) that led to an application for a system that transmitted to the customer at a high data-rate for supporting entertainment features such as video on demand or simple download of any desired data. The emerging market ultimately led to the conception of the Asymmetric Digital Subscriber Line (ADSL). ADSL allows for simultaneous transmission of digital data and Plain-Old Telephone Service (POTS) signal on a twisted-pair wire. This fact turned out to be one of the success factors during the commercial launch of this product since it could be easily put on top of the already widely spreaded POTS system.

R. Gaggl, *Delta-Sigma A/D-Converters*, Springer Series in Advanced Microelectronics 39,
DOI 10.1007/978-3-642-34543-2_1, © Springer-Verlag Berlin Heidelberg 2013

Beside the use of electrical conductors the wireless technology has become a very important branch in telecommunications for transferring any kind of data over a short or long distance. The wireless technology has been the driving force for enabling new services such as satellite television, mobile telephones, personal digital assistants and wireless networking. For the home and business user, wireless has become popular due to ease of installation, and location freedom with the gaining popularity of laptops. The data exchange between computers demands for broadband and flexible computer network solutions such as in a local area network (LAN). Basically, a LAN covers a small geographic area, like a home, office, or group of buildings and provides access to a computer network at high data transfer rates. A Wireless Local-Area Network (WLAN) is a local-area network (LAN) without employing any kind of wires. WLAN solutions have been around for more than a decade, but these are just beginning to gain momentum because of falling costs and improved standards. In 1991, the IEEE 802.11 committee had started its activities to develop a standard for wireless LANs. By 1996, the technology became relatively mature, a variety of applications had been identified and addressed. The IEEE 802.11 standard supports various versions such as 802.11a and 802.11g offering 54 Mbit/s in the 5 GHz band and the 2.4 GHz, respectively.

Regardless of wireline or wireless communication solutions, on the physical layer there is a strong demand for wide band and high performance Analog-to-Digital (A/D)-data-converters being the link between the analog and digital world. Providing digitized data with sufficient quality allows for doing digital signal processing of the received data stream. The required performance of the A/D-converter is influenced by architectural considerations on system level in order to not compromise the maximum data-rate. That means, beside the right approach on system level, the architectural choice for the A/D-data-converter implementation is crucial for achieving the targeted performance and for keeping the overall costs low of the whole system implementation.

1.2 Organization of This Book

This book deals with oversampling converters processing baseband signals, i.e. signals with spectra centered around zero Hertz Direct Current (DC). The focus is on the design of delta-sigma modulators embedded in digital CMOS technologies for communication systems. The target applications will be Asymmetric Digital Subscriber Line (ADSL) as well as Wireless Local-Area Network (WLAN) covering bandwidths from 300 kHz up to 20 MHz. The resolution of the different converter solutions results in 10 bit to 14 bit. Chapter 1 gives a brief overview about the architecture of an ADSL and WLAN system.

Chapter 2 covers the basic aspects about analog-to-digital conversion as well as the main limitations on delta-sigma modulation techniques. The focus will be on embedded delta-sigma modulators discussing single-loop and cascaded-loop topologies as well as switched-capacitor and time-continuous implementations.

A design methodology is presented for continuous-time delta-sigma converters. An introduction on pulse-width modulation is given and the impact on clock-jitter will be shown. The further outline of this book is structured into three main parts.

Chapter 3 will present a delta-sigma converter designed for ADSL in 180 nm CMOS employing a dynamic biasing technique. Considerations are given on architectural-level as well as on circuit-design followed by measurement results. Considerations on the SNR are done by finding a careful balance of all relevant noise contributors. The power-optimization is addressed on circuit level by introducing a dynamic-biasing of all operational-amplifiers. An area- and power-efficient realization of a single-loop modulator consisting of a 2nd-order loopfilter and a 3 bit quantizer with an oversampling-ratio of 96 is presented. The delta-sigma modulator features an 85 dB dynamic-range (DR) over a 300 kHz signal bandwidth. The measured power consumption of the ADC core is 15 mW only from a mixed 1.8 V and 3.3 V supply. An innovative biasing circuitry is introduced for the switched-capacitor integrators. The FOM taking the dynamic-range as reference results in $1.8 \frac{\text{pJ}}{\text{conv}}$.

A feed-forward delta-sigma converter is described in Chap. 4 presenting the architecture on system-level and showing solutions on circuit-level embedded in 130 nm digital CMOS. Attention is paid on system-level by going for a feedforward topology in order to improve the SNR by increasing the tolerable signal swing. Implementation issues are discussed for an operational-amplifier, an integrator, a quantizer and a reference buffer in open-loop configuration. A passive switched-capacitor adder is introduced to keep the power drain low. A 2nd-order modulator with high oversampling-ratio (OSR = 192) is presented employing a switched-capacitor loopfilter and a 3 bit quantizer. The delta-sigma modulator features a 14 bit and 13 bit dynamic-range over a 276 kHz and 1.1 MHz signal bandwidth, respectively. The measured power consumption of the ADC-core is 8 mW only. Including an on-chip reference buffer the total power consumption results in 15 mW from a single 1.5 V supply. The FOM of the ADC-core results in $0.7 \frac{\text{pJ}}{\text{conv}}$.

In Chap. 5, a broadband delta-sigma A/D-converter is presented for a Wireless Local-Area Network (WLAN) application. The converter is implemented in a 65 nm CMOS technology. An area- and power-efficient realization of a single-loop delta-sigma modulator consisting of a 3rd-order continuous-time loopfilter and a time-encoding quantizer (TEQ) is discussed. The introduced TEQ will be implemented inside a delta-sigma modulator by replacing a multibit quantizer. An innovative TEQ is introduced to overcome design issues in a 1.0 V supply-voltage 65 nm digital CMOS technology. The TEQ allows an exchange of *amplitude-resolution* by *time-resolution*. The approach of time-resolution alleviates the scaling difficulties of mixed-signal circuits in nanometer technologies. This is accomplished by replacing a flash A/D-converter by a modulated oscillator producing a two-level signal. The TEQ comprises a self-oscillating pulse-width modulator (PWM) embodied as a passive circuit keeping the power-drain small. A clocked comparator is placed within the PWM, hence, the duty-cycle of the rectangular signal is discrete in time. The clocked PWM produces a binary rectangular signal with a period which is $8\times$ longer than one clock-period. The duty-cycle is modulated by the output of the loopfilter and represents the information of a multibit signal. The binary

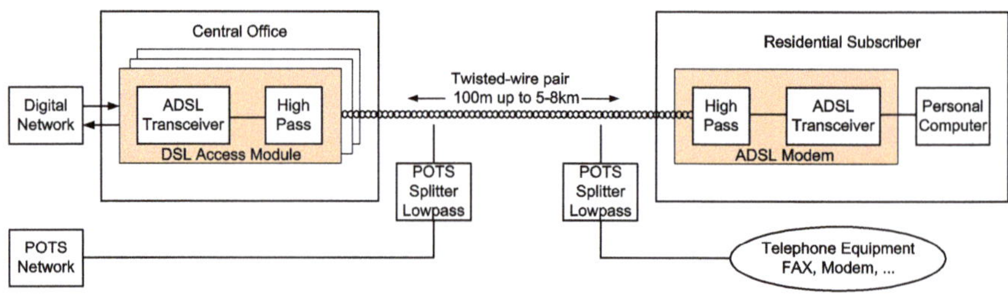

Fig. 1.1 Digital access through a POTS-splitter

signal is fed back through a single-bit D/A-converter to close the delta-sigma loop. The innovative step lies in the fact that a binary circuitry provides a performance of a multibit solution by employing resolution in time. The measured power consumption of the total ADC is 7 mW only thanks to the single-bit architecture of the TEQ and the D/A-converter. The delta-sigma modulator features a 63 dB dynamic-range over a 20 MHz signal bandwidth supporting the IEEE standard 802.11n (see Table 1.2). Employing the high-speed devices by choosing a clock-frequency of 2.5 GHz results in an attractively small chip-area of 0.079 mm^2 and in a FOM of 0.15 pJ/conversion-step as described in (2.7). Furthermore, a detailed concept study is presented for a TEQ showing architectural considerations as well as main issues on circuit implementation in a nanometer CMOS-technology (65 nm). Comprehensive measurement results will conclude chapter five.

The overall conclusions (Chap. 6) will summarize the issues on analog design in different CMOS technology nodes optimized for digital circuits and compare different solutions by calculating the achieved Figure Of Merit (FOM). The outlook on future activities will be discussed as well. At the end references are listed for all relevant publications in the field of this book.

1.3 Asymmetric Digital Subscriber Line

Basically, the Public Switched Telephone Network (PSTN) is dominated by digital technologies. High-speed digital transmission links between different telephone Central Office (CO) are capable of multiplexing and concentrating local data traffic according to the digital hierarchy. By bypassing the POTS interface at a local CO, a DSL system can utilize the full potential of a copper twisted-pair telephone subscriber loop to deliver a transmission throughput of up to a few Mbits per second depending on the length and the quality of the used line [73].

Figure 1.1 shows the possibility of providing digital transmission through the addition of POTS splitters. The shown configuration for ADSL employs a POTS splitter for separation of voice and data through highpass and lowpass filters to provide co-existing services.

The major network elements are the DSL access module in the CO and the ADSL modem at the residential subscriber. At both ends of the telephone twisted-pair,

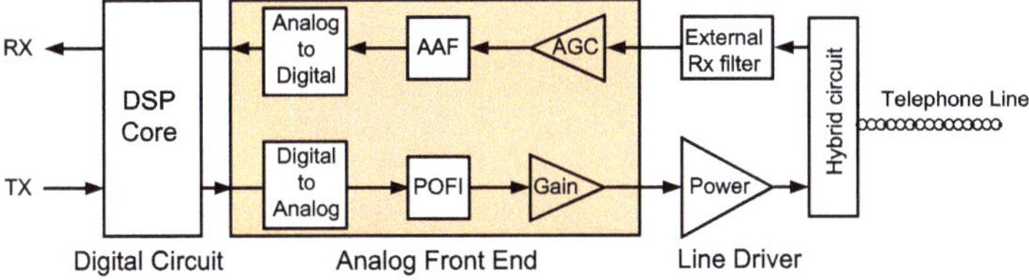

Fig. 1.2 A block diagram of an ADSL modem

a splitter-combiner filter multiplexes the POTS signals with the ADSL signals. This multiplexing can be accomplished by highpass- and lowpass-filtering since ADSL transmits in frequency overlay, that is, at frequencies above the existing telephony band as discussed in the next subsection. These filters also prevent mutual interference of both signals [73].

ADSL is currently a standard application in the field of wired communications. The data-rates for internet access at multiples of 1.5 Mbit/s up to 24 Mbit/s and a narrow upstream up to 3.5 Mbit/s (from the end user to the network) command channel could easily be adapted. The ADSL standard offers a flexible setting of data-rates and allocation of frequency bands. Rather than being limited to a fixed bit-rate, ADSL can be tuned dynamically to deliver any predefined rate, or even the maximum available bit-rate on the telephone line. With Asymmetric Digital Subscriber Line 2+ (ADSL2+), a new standard for increased data-rates and improved loop reach has become available. To remain competitive in this aggressive market segment, it is necessary to fulfill all the standard requirements and also optimize the performance of the whole product and its power consumption. Especially in the CO and Digital Loop Carrier (DLC) applications, the power consumption is a major issue. Such system solutions are based on high-density designs with typically 64 channels per linecard or even a higher number can be supported.

The Analog Front End (AFE) for an ADSL transceiver will be described briefly. The AFE serves as a four-wire system approach comprising a transmit-path and a receive-path. Figure 1.2 shows one single channel of a transceiver chip set comprising a line-driver, the AFE and a digital signal-processor. Furthermore, external components are used for building up a hybrid circuit as well as several filter stages and a transformer.

The hybrid circuit serves as four-wire to two-wire translator, since the telephone line supports two wires only. The transmit output of the AFE is connected to the line-driver which supplies the appropriate signal level to the line. The analog lowpass filter block in the transmit-path (POFI) serves as a smoothing filter after the Digital-to-Analog (D/A) conversion. The following gain stage drives the line-driver input impedance providing the required signal level to maintain the transmitted signal power on the line. The analog input signal to the AFE is received via the hybrid circuit, which is one of the main component influencing the overall system performance. The receive signal loss due to the hybrid has to be kept as small as possible. The automatic gain-control stage (AGC) amplifies the small receive signal accordingly and the anti-aliasing filter prevents any undesired folding of out of band

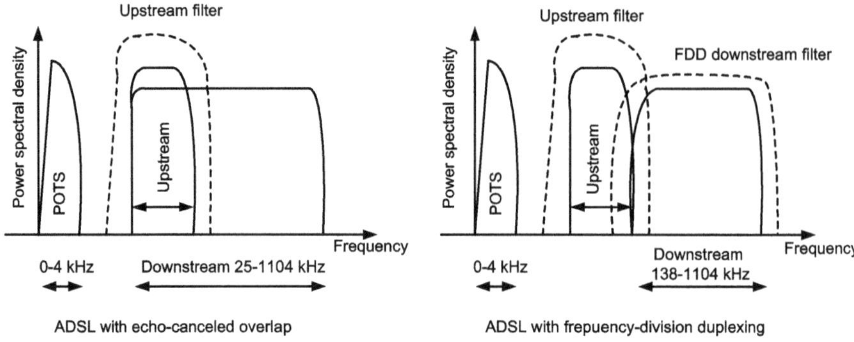

Fig. 1.3 Frequency band usage in ADSL, EC and FDM

disturbers into the band of interest before the signal enters the A/D-converter. An important transfer function is the so called *echo-path*. The echo-path describes the desired attenuation of the transmit signal leaking back through the hybrid circuit to the receive signal. Especially on long loops, the remaining echo is the dominating signal in the receive-path. The echo attenuation depends on the connected line and how well the hybrid circuit has been adapted to the attached line. This means, it is crucial to take the occurring echo into account in all system considerations.

1.3.1 ADSL System Configuration

In order to provide a data exchange in both directions at the same time, two duplexing variants have been standardized. The first one, depicted in Fig. 1.3 (shown left), has a downstream signal overlapping the upstream and requires echo canceling to separate both signals. The second one uses frequency-division duplexing (FDD) with separate bands for the up- and down-stream, as shown right in Fig. 1.3. Obviously, the echo-cancellation implementation is much more difficult, since the frequency-division duplexing can be made with filters and requires a much smaller dynamic-range of the analog components. This fact influences strongly the robustness and gave an early preference for the second variant.

The upstream band is used for communication from the Customer Premises Equipment (CPE) to the telephone central-office. The downstream band is used for data transmission from the central-office to the end user. Each of these bands is further divided into smaller frequency channels of 4.3125 kHz. During initial training, the ADSL modem tests which of the available channels have an acceptable signal-to-noise ratio. The distance from the CO, noise on the twisted-pair, or interference from any radio stations may introduce errors on some frequencies. By keeping the channels small, an error on one frequency thus need not render the line unusable. The channel will not be used, merely resulting in reduced throughput on an otherwise functional ADSL connection. The discrete-multi-tones

Table 1.1 Overview of ADSL standards

Standard name	Common name	Downstream rate	Upstream rate
ANSI T1.413-1998	ADSL	8 Mbit/s	1.0 Mbit/s
ITU G.992.1	ADSL (G.DMT)	8 Mbit/s	1.0 Mbit/s
ITU G.992.1 Annex-A	ADSL over POTS	8 Mbit/s	1.0 Mbit/s
ITU G.992.1 Annex-B	ADSL over ISDN	8 Mbit/s	1.0 Mbit/s
ITU G.992.2	ADSL-Lite (G.Lite)	1.5 Mbit/s	0.5 Mbit/s
ITU G.992.3/4	ADSL2	12 Mbit/s	1.0 Mbit/s
ITU G.992.3/4 Annex-J	ADSL2	12 Mbit/s	3.5 Mbit/s
ITU G.992.3/4 Annex-L	RE-ADSL2	5 Mbit/s	0.8 Mbit/s
ITU G.992.5	ADSL2+	24 Mbit/s	1.0 Mbit/s
ITU G.992.5 Annex-L	RE-ADSL2+	24 Mbit/s	1.0 Mbit/s
ITU G.992.5 Annex-M	ADSL2+M	24 Mbit/s	3.5 Mbit/s

(DMT) modulation technique was chosen for the first International Telecommunications Union-Standardization Sector (ITU-T) ADSL standard. Table 1.1 depicts an overview about the different ADSL standards. Note that the displayed data-rates are theoretical maximums. The commonly deployed annex-A splits both frequency bands at 138 kHz. Whereas annexes-J and -M shift the upstream and downstream frequency-split up to 276 kHz in order to boost upstream rates. Additionally, the "all-digital-loop" variants of ADSL2 and ADSL2+ (annexes-I and -J) support an extra 256 kbit/s of upstream if the bandwidth normally used for POTS voice calls is allocated for ADSL usage. While the ADSL access utilizes the 1.1 MHz band, ADSL2+ utilizes the 2.2 MHz band.

1.3.2 A/D-Converter Requirements for ADSL

It has been a common practice to combine the current and upcoming expected capabilities of Very Large-Scale Integrated Circuits (VLSI) with the desired performance of a DSL system under development. This combination has resulted in both a challenge and rewarding experience for DSL concept engineers. For example, high resolution, low sampling-rate and low resolution, high sampling-rate A/D-converters have been developed for voice and video applications, respectively. The development of ADSL technology resulted in a market need of high resolution and high sampling-rate, 13 bits and more than 1 MHz. Due to the asymmetrical nature of ADSL, that means different bandwidth for upstream and downstream, the band of interest for the A/D-converter depends on which side of the twisted-pair the converter will be placed. In other words, the band of interest results at the central-office's side and at the subscriber's side in 276 kHz and 1.1 MHz (2.2 MHz), respectively. The dynamic range depends on the chosen system architecture such as the hybrid and comes to the range of 13 bits to 14 bits. In order to not compromise the maximum data-rate the harmonic distortion must not exceed 80 dBc for a single tone. Finally, the intermodulation products of the applied discrete multi-tones count for the linearity and will be expressed in Missing-Tone Power Ratio (MTPR). An area- and power-efficient realization of a converter is mandatory to remain competitive in the market. This point will be one of the major challenges in the design procedure beside keeping the required performance. The right choice for the converter topology and architecture will be the success factor for the product and needs to be done very carefully.

1.4 Wireless Local-Area Network

Wireless Local-Area Network (WLAN) solutions transfer data through the air using radio frequencies instead of cables providing location independent network-access between computing devices. In the corporate enterprise, WLAN solutions are typically used as the final link between the existing wired network and a group of client computers, giving these users wireless access to the full resources and services of the corporate network across a building. The major benefit and motivation from WLAN is increased mobility. In contrast to conventional wired network connections, users can move about almost without restriction and access Local-Area Network (LAN) systems from nearly anywhere. Furthermore, a WLAN approach offers cost-effective network setup for locations which are hard to wire such as older buildings and solid-wall structures. Costs per device and user can be reduced in case of frequent modifications in dynamic environments for the wiring and installation [49].

The 802.11 specification as a standard for WLAN solutions was ratified by the Institute of Electrical and Electronics Engineers (IEEE) in the year 1997. This version of 802.11 provides for 1 Mbps and 2 Mbps data-rates and a set of fundamental signaling methods and other services. Like all IEEE-802 standards, the 802.11 standards focus on the bottom two levels of the Open System Interconnection (OSI) model, the physical layer and link layer (see Fig. 1.4).

Any LAN application, network operating-system, protocol, including Transmission Control Protocol/Internet Protocol (TCP/IP), will run on a an 802.11-compliant WLAN solution as easily as they run over some Ethernet application. The WLAN architecture is configurable in a quite flexible manner. Each computer, mobile, portable or fixed, is referred to as a station in WLAN. The difference between a portable and mobile station is that a portable station moves from point to point but is only used as a fixed point. Mobile stations can even access the LAN during movement. When two or more stations come together to communicate with each other, they form a Basic Service Set (BSS). The minimum BSS consists of two stations. 802.11 LANs use the BSS as the standard building block. A BSS that stands alone and is not connected to a base is referred to as an ad-hoc network or a so called Independent-BSS. That means, all stations communicate only peer to peer without any base and no one gives permission to talk. Such a network can be set up rapidly. When BSS's are interconnected the network becomes one with infrastructure comprising two or more BSS's interconnected to a so called distribution system (DS).

Fig. 1.4 OSI model for
WLAN

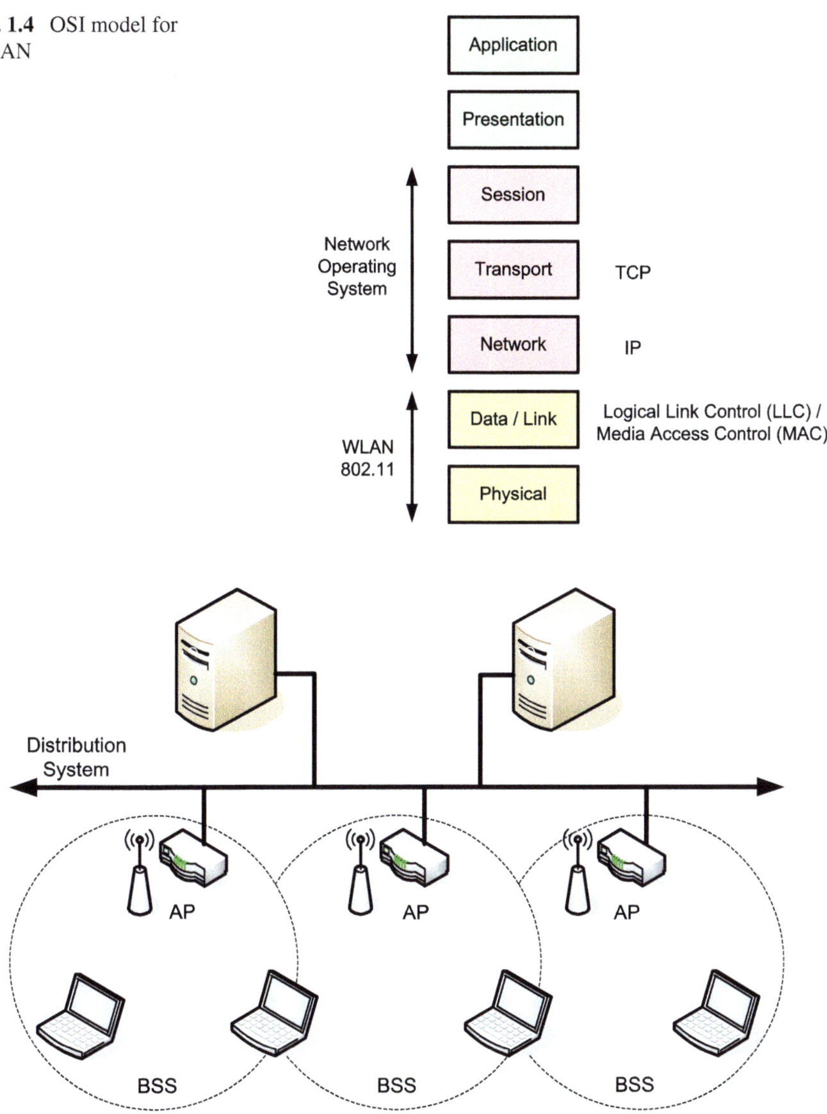

Fig. 1.5 WLAN infrastructure mode

The implementation of the DS is not specified by IEEE-802 standards. Therefore, a distribution system may be created from existing or new network technologies. Entry to the distribution system is accomplished with the use of access points (AP), which are addressable and offer distribution system services along with bridging functionality.

Data moves between the BSS and the distribution system with the help of these access points (AP), as depicted in Fig. 1.5 [49].

Table 1.2 Overview of 802.11 PHY-layer

Standard	Data rate per stream	Transfer method	Frequency band
IEEE 802.11	1 and 2 Mbit/s	optical	infrared
IEEE 802.11	1 and 2 Mbit/s	FHSS	2.4 GHz
IEEE 802.11	1 and 2 Mbit/s	DSSS	2.4 GHz
IEEE 802.11a	6, 9, 12, 18, 24, 36 and 54 Mbit/s	OFDM	5 GHz
IEEE 802.11b	5.5 and 11 Mbit/s	DSSS	2.4 GHz
IEEE 802.11g	6, 9, 12, 18, 24, 36, 48 and 54 Mbit/s	OFDM	2.4 GHz
IEEE 802.11n	7.2 up to 150 Mbit/s	OFDM	2.4 GHz/5 GHz
IEEE 802.11ac (draft)	433 up to 867 Mbit/s	OFDM	5 GHz

The physical layer is the bottom layer of the OSI model (see Fig. 1.4). The three different physical layers originally defined in 802.11 included two spread-spectrum radio techniques and a diffuse infrared specification. The spread-spectrum technique spreads the narrow baseband signal over a wide frequency band in order to improve the immunity against any discrete disturbers. The originally radio-based standards operate within the 2.4 GHz band. This band is called Industrial-Scientific-Medical (ISM) band which does not require any licensing. Spread-spectrum techniques increase reliability, boost throughput and allow many unrelated products to share the spectrum without explicit cooperation and with minimal interference. The original 802.11 wireless standard defines data-rates of 1 Mbps and 2 Mbps via radio waves using frequency hopping spread spectrum (FHSS) or direct sequence spread spectrum (DSSS). It is important to note that both techniques are fundamentally different signaling mechanisms and will not interoperate with each other. Using the frequency hopping technique, the 2.4 GHz band is divided into 75 adjacent 1 MHz subchannels. The sender and the receiver agree on a hopping pattern and data is sent over a sequence of the subchannels. Each conversation within the network occurs over a different pattern to avoid using the same subchannel by two senders simultaneously. In contrast, the direct sequence spread spectrum technique divides the 2.4 GHz band into 14 adjacent 22 MHz subchannels. Data is sent across one of these 22 MHz subchannels without hopping to other channels [49].

The most critical issue affecting WLAN demand has been limited throughput. Recognizing the critical need to support higher data-rates, new standards were ratified as shown in Table 1.2. The transfer method employs three different principles of modulation schemes, Frequency-Hopping Spread-Spectrum (FHSS) or Direct-Sequence Spread-Spectrum (DSSS) or Orthogonal Frequency-Division Multiplexing (OFDM). Basic information about these methods can be found in [49].

In Table 1.2 an overview about the different WLAN standards is shown. The different standards are listed with the achievable data-rates and the used transfer methods along with used frequency bands. The first generation of IEEE 802.11 was released in 1997 followed by the second generation 802.11b two years later. The main use case was the internet and the exchange of e-mails. The third generation 802.11g was driven by the demand for enabling the rich-data internet experience such as sharing pictures and low-resolution video streaming. In 2007, the increasing demand for higher data-rates consequently asked for the fourth generation WLAN 802.11n. IEEE 802.11n is an amendment which improves upon the previous 802.11 standards by adding multiple-input multiple-output antennas (MIMO). The most common data-rate is 150 Mbit/s and by employing 4 streams in parallel 600 Mbit/s are achievable. 802.11n operates on both the 2.4 GHz and the less frequently used 5 GHz bands. The standard for the fifth generation has not yet being released fully. The IEEE 802.11ac standard is still under development which will provide high throughput in the 5 GHz band. This specification will enable multi-station WLAN throughput of 1000 Mbit/s and a maximum single link data-rate of at least 433 Mbit/s, which is at least three times faster than that of most commonly used wireless standard 802.11n. The increased speed can be accomplished by using wider RF bandwidth (80–160 MHz) and up to 8 streams. The fifth generation will bring fast, high quality video streaming and fast data syncing and backing up of data storage to notebooks, tablets and mobile phones.

The data-link layer is the second layer of the OSI model (see Fig. 1.4). The data-link layer within 802.11 consists of two sublayers: Logical Link Control (LLC) and Media Access Control (MAC). The MAC is designed to support multiple users on a shared medium by having the sender sensing the medium before accessing it. In an 802.11 WLAN, collision detection is not possible as for conventional Ethernet which is known as the "near/far" problem: to detect a collision, a station must be able to transmit and listen at the same time, but in radio systems the transmission drowns out the ability of the station to "hear" a collision. To overcome this issue, 802.11 uses a protocol known as Carrier-Sense Multiple-Access with Collision-Avoidance (CSMA/CA). CSMA/CA attempts to avoid collisions by using explicit packet acknowledgment, which means an acknowledgment packet is sent by the receiving station to confirm that the data packet arrived intact. In other words, this approach provides a way of sharing access over the air. Finally, the MAC-layer provides robustness features such as the checksum for cyclic redundancy check and packet fragmentation and reassembling of received packets. Time-bounded data such as voice and video is supported in the MAC specification through the point coordination function where a single access-point controls access to the media and the access time is spliced to guarantee an acceptable latency. It is worth mentioning that beside WLAN further wireless solutions have been standardized such as Hiper-LAN, Home-RF, Bluetooth Zig-Bee and WiMax [49].

1.4.1 WLAN 802.11n System Configuration

The 4th generation of WiFi networking standards IEEE 802.11n was released in 2009. This release is an amendment which improves upon the previous 802.11 standards by adding multiple-input multiple-output antennas (MIMO). Up to a number of 4 MIMO streams are supported in parallel. The IEEE has approved the amendment and it was published in October 2009. This extension defines a new physical layer supporting data-rates up to 150 Mbit/s and optionally up to 600 Mbit/s

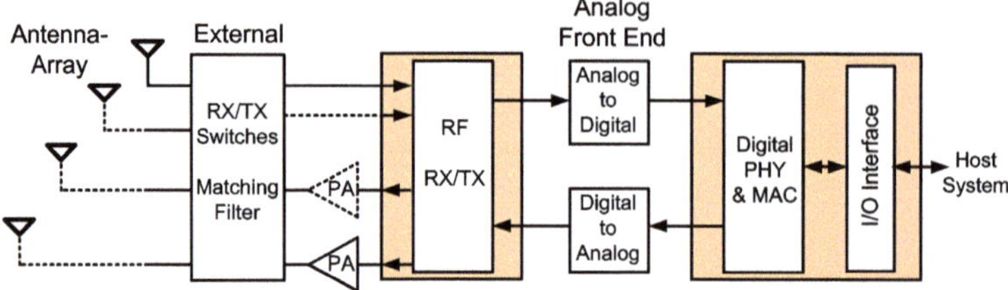

Fig. 1.6 WLAN block diagram

by exploiting 4 MIMO streams. Furthermore, the MAC-layer is extended to ensure high efficiency at high data-rates along with an improved offering for feasibil-
ity. The core of the physical layer is an intelligent antenna system—the so called
smart-antenna approach. The smart antennas comprise several identical antennas
with a distance of $\frac{\lambda}{2}$ whereas the antennas are driven by an intelligent algorithm for
signal-processing. This intelligent controlling improves the quality of receiving and
transmitting data. This solution can achieve higher data-rates. Applying the smart-
antenna approach on both sides, receiver and transmitter, results in the so called
Multiple-Input, Multiple-Output (MIMO) system. The basic idea of MIMO is the
fact that several antennas can receive more signals and can transmit more signals
than a single antenna, whereby the data-rate can be improved. The use of several
antennas on both sides is an already well known approach in wireless systems [49].

A basic block diagram for a WLAN solution is depicted in Fig. 1.6. The data
are received and transmitted through an array of antennas. The number of antennas is configurable due to the supported standard of 802.11. The employment of
semi-duplex transmission mode is one important feature of WLAN. That means,
the whole system is either in receive mode or in transmit mode. Data transmission
in both directions at the same instant is not possible. This fact allows an easier circuit implementation around the antenna. Time multiplexing between receive and
transmit direction is employed by means of a switch in front of the antenna.

The antenna receives a signal which goes through a matched filter and is further switched to an additional lowpass filter stage. Moreover, the signal enters the
RX-path of the RF-transceiver for down-conversion into the baseband. Several techniques are well known such as heterodyne- or homodyne-receivers [5]. After removing any occurring dc-offset, the RF-signal is directly down-converted by means of
a mixer and an appropriate local oscillator frequency. The demodulated signal is
applied to a programmable gain amplifier and followed by the required anti-aliasing
filter. The baseband signal needs to be digitized by means of an A/D-converter to
provide digitized receiver signal to be processed in a digital signal processing unit.
The D/A-converter provides an analog signal defined in the digital signal processing unit for the transmit-path. After passing through the smoothing filter stage a
low-noise-amplifier (LNA) drives the amplified signal into the mixer to up-convert
the transmit signal into the desired frequency band. The baseband signal is modulated and up-converted by the TX-path of the RF-transceiver. Any occurring offset
voltage is suppressed in the channel, the RF signal is amplified accordingly and

switched to the antenna. The power amplifier (PA) provides sufficient power to the antenna. The multiplexing switch connects the power amplifier to the antenna array. A common RF-transceiver consists of several data channels whereas one data channel comprises both I/Q sub-channels featuring quadrature modulation [5].

1.4.2 A/D-Converter Requirements for WLAN 802.11n

The A/D-converter connects the RF-transceiver to the digital signal-processor as illustrated in Fig. 1.6. Commonly, different technology nodes are used for the RF-transceiver and the digital part to optimize the overall production costs. The choice for A/D-converter placement either onto the RF-die or on the baseband-die needs to be done carefully. One major impact is the required type of interface. A converter on the RF-die implies an high-speed digital interface to reduce the number of package pins. High switching activities may cause severe crosstalk on the RF-die. This fact favors the converter placement onto the digital chip to allow an analog interface avoiding steep transitions. This latter approach constrains the technology node for the converter. In our days, cost-effective implementations go for nanometer technologies in case of digitally dominated integrated-circuits (IC). In other words, the A/D-converter for WLAN needs to be designed in a nanometer technology. The development of WLAN standards has resulted in a need of moderate resolution but high-speed A/D-converters, that means 10 bits and up to 20 MHz. The total data-rate will be deteriorated by any occurring mismatch between both I/Q-channels, since quadrature-modulation is employed. To overcome that issue, the tolerable I/Q-mismatch needs to be specified in terms of I/Q-gain-error and I/Q-phase-error. Proper system simulations revealed a specification of 0.5 dB and $5°$, respectively. The group delay variation must not exceed 10 ns. The transition-time between sleep- and active-mode needs to be faster than 1 μs to allow power saving during multiplexing in time domain of transmit- and receive-path. An area- and power-efficient converter realization is mandatory to remain competitive in the market. This point will be one of the major challenges in the design procedure beside keeping the required performance in a challenging nanometer technology.

Chapter 2
Limitations of Delta-Sigma Converters

This chapter gives an brief overview of A/D-conversion focusing on delta-sigma modulators. For a comprehensive analysis of delta-sigma converters the reader is referred to later mentioned references. The sampling and quantization of analog signals is briefly discussed. The main converter specification items are introduced. Next, the principle of oversampling and delta-sigma conversion is covered followed by some architectural aspects and implementation approaches. Pulse-width modulation will be introduced based on a delta-sigma modulator for low voltage applications. Finally, the impact on jitter for A/D-conversion is briefly discussed.

2.1 Basics of A/D-Conversion

The conversion of analog signals to the digital domain can be divided into two basic operations: uniform sampling in time and quantization in amplitude. Generally, the analog signal can be assumed to be limited within a arbitrary frequency band, i.e. $|f_{sig}| \leq f_B$, where f_B is defined as the analog signal bandwidth. Under this assumption, the sampling in time is a completely invertible process. The sampling process can be modeled as a multiplication with a sequence of Dirac-pulses by the analog signal. A multiplication in time domain results in a convolution in frequency domain due to the properties of *Laplace transformation*. In other words, a quantization in time can be considered as a periodization in frequency, which is illustrated in Fig. 2.1. There, the considered analog signal is sampled at uniform time intervals T_S, the so called sampling time, or with a fixed frequency f_S, resulting in a periodicity of the original analog spectrum at multiples of f_S. The periodization itself comes from the sampling process considered as a multiplication of the analog signal with a train of Dirac-pulses spaced at T_S in time. In frequency domain the sampling process can be considered as a convolution of the analog spectrum with a periodic train of Dirac-pulses spaced at f_S. The sampling frequency can be expressed as follows: $f_S = \frac{1}{T_S}$.

R. Gaggl, *Delta-Sigma A/D-Converters*, Springer Series in Advanced Microelectronics 39,   15
DOI 10.1007/978-3-642-34543-2_2, © Springer-Verlag Berlin Heidelberg 2013

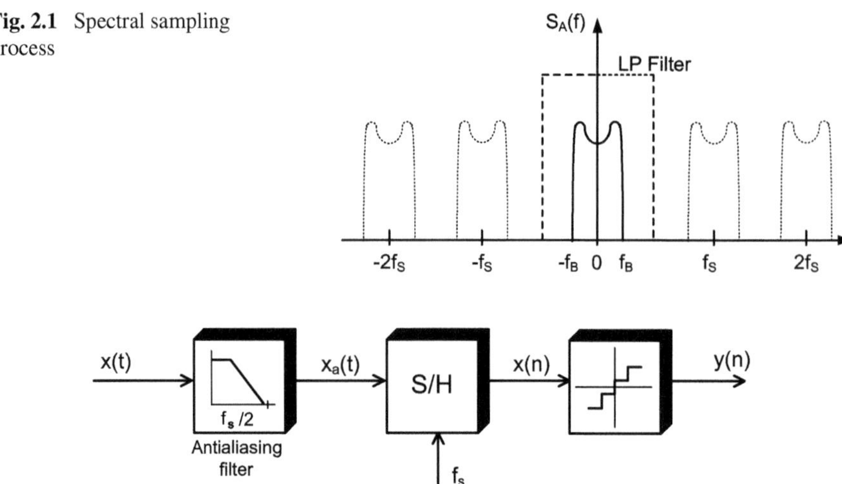

Fig. 2.1 Spectral sampling process

Fig. 2.2 Principle of an A/D conversion

As shown in Fig. 2.1 it can be easily seen that by simple lowpass filtering, the original analog signal can be reconstructed. This fact is true as long as the sampling itself does not result in overlap or aliased regions. This is fulfilled if:

$$f_S \geqq 2f_B = f_N, \tag{2.1}$$

which is known as the *Nyquist theorem*, where f_N is the Nyquist frequency. In real implementations, a so called anti-aliasing filter is used preceding the sampling process. This enforces the Nyquist theorem and results in a proper sampling operation. Therewith, the basic A/D-operation can be split into three basic parts, an anti-aliasing filter, a Sample-and-Hold (S/H) stage and a quantizer as shown in Fig. 2.2. Notice, the signal $x(n)$ after the sampling process is discrete in time and still continuous in amplitude whereas the signal $y(n)$ after the quantizing process is discrete in time and amplitude. This process is called amplitude-encoding in contrast to a time-encoding approach, which will be discussed later in this chapter.

Assuming a brickwall lowpass filter for cutting off the unwanted high frequency signals, the S/H-stage can be clocked at Nyquist frequency, which is called a *Nyquist-Rate converter*. It is obvious that a brickwall filter with zero transition band is absolutely impractical for real implementations. Analog filters with a smooth roll-off in their transition band are a better approach to result in a less costly solution in terms of smaller chip area and less power drain. Therefore, many A/D-converters operate at sampling frequencies higher than f_N, which yields to following definition:

$$OSR = \frac{f_S}{f_N} \tag{2.2}$$

as the Over-Sampling Ratio (OSR) of the A/D converter.

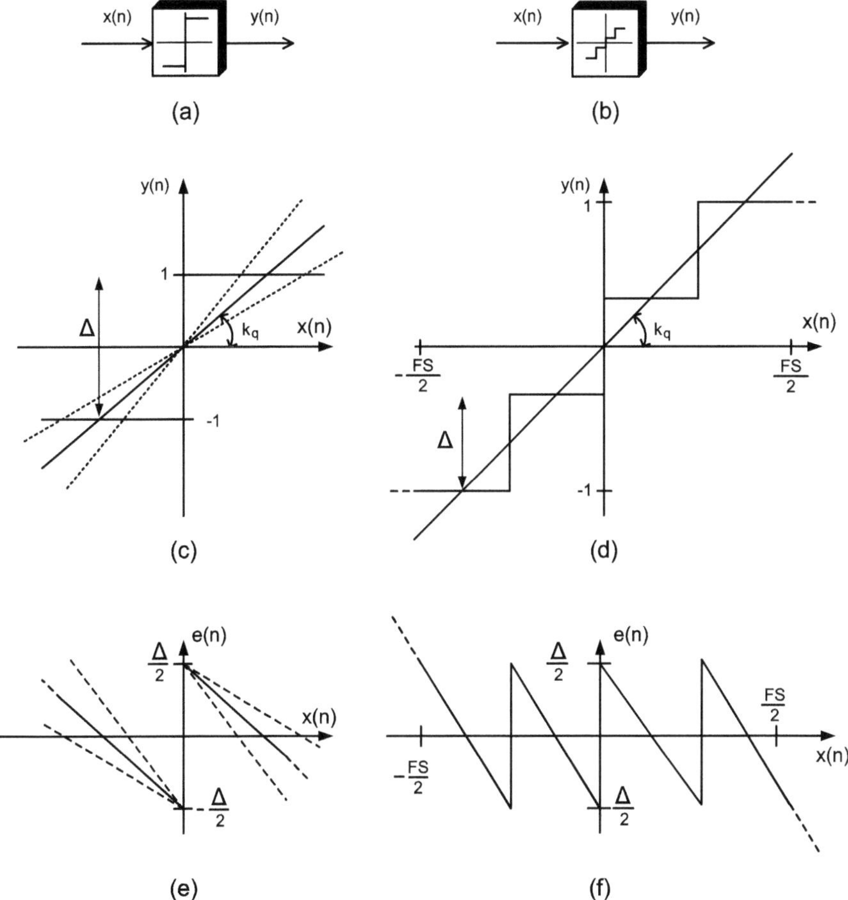

Fig. 2.3 Quantizer transfer curves and quantization error

The quantization is a process of quantization in amplitude encoding a continuous range of analog values into a set of discrete levels. In other words, an infinite number of input amplitudes $x(t)$ are mapped into a finite number of output amplitude values $y(n)$ which clearly leads to a noninvertible operation. Thus, even an ideal quantizer inherently introduces errors to the output signal $y(n)$. Therefore, the primary objective in A/D-converter design is to limit exactly this error. The quantizer can be characterized by its number of different output levels. A quantizer having a number of bits B can map the analog input amplitudes into 2^B discrete levels. In other words, the quantizer shows B-bits of resolution. The uniform spaces between two adjacent quantizer output levels are defined as the quantizer step width:

$$\Delta = \frac{FS}{2^B - 1},\tag{2.3}$$

where FS is the full-scale output range. This is shown in Fig. 2.3, where a single-bit (a) and multi-bit (b) quantizer are given with their input-output characteristic (c) and (d). As the quantizer input signal $x(n)$ rises from $-FS/2$ up to $+FS/2$ the output signal $y(n)$ is quantized to one of the 2^B different output levels. The resulting quantization error is defined as the difference of the quantizer input and its output, which is shown in Fig. 2.3(e) and (f). Obviously, an input signal exceeding the valid input range $[-FS/2\ldots+FS/2]$, results in an increasing quantization error, which is commonly known as the overload or saturation region of a quantizer. The indication of slope k_q stands for the non-unity gain of a quantizer. Since the input and output range of the quantizer are not necessarily equal. In particular, the gain of a single-bit quantizer is not defined exactly, but can be chosen arbitrarily, as illustrated in Fig. 2.3(c). This is because the output of the single-bit quantizer only depends on the polarity but not on the amplitude of the input signal.

In the open literature a *white noise model* for the quantizer has been derived. This model is valid under the assumption that the input changes rapidly from sample to sample by amounts larger than the step size Δ. Then the quantization error appears uncorrelated from sample to sample and has equal probability of being anywhere in the range of $[\frac{-\Delta}{2}, \frac{+\Delta}{2}]$, provided that no overload occurs. This observation leads to the assumption of having statistical properties for the quantization error and makes the basis of the *additive white noise model* of the quantizer. The white noise approximation drastically simplifies the analysis of an A/D-converter solution, because a deterministic nonlinear system is replaced by a stochastic linear one. That means, the quantization error becomes a quantization noise. In a real converter implementation, this does not strictly hold true, but nevertheless the quantizer's white noise model has been yielding good results. It is worth mentioning that the total quantization error power is independent of the sampling frequency and is only determined by the quantizer resolution. Since the output signals at the quantizer are discrete in time, all the quantization noise power is folded into the frequency range $[\frac{-f_S}{2}\ldots\frac{+f_S}{2}]$. Thus, with the white noise approximation the power spectral density of the quantization noise can be derived and yields as follows:

$$S_q = \frac{\Delta^2}{12}\frac{1}{f_S}. \tag{2.4}$$

This white noise model gives a sufficiently good approximation for the quantization process of an A/D conversion.

2.2 Main A/D-Converter Specifications

The most important specifications used in this book are briefly discussed. Basically, the characteristics of A/D-converters can be classified into two groups: static and dynamic performance metrics. The first include monotonicity, offset, gain error, Differential Non-Linearity (DNL) and Integral Non-Linearity (INL). The dynamic performance can be expressed by Signal-to-Noise Ratio (SNR), Total Harmonic

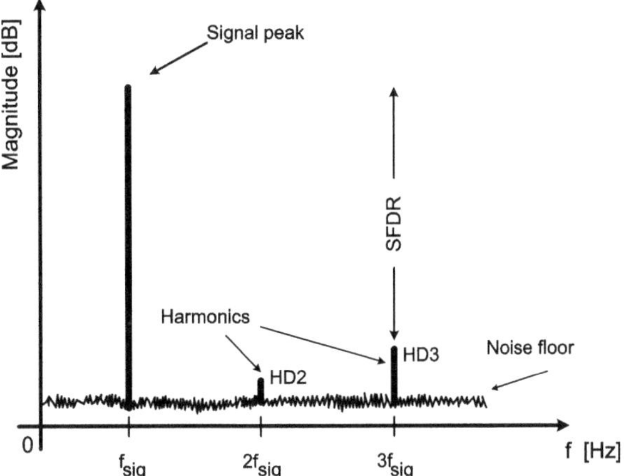

Fig. 2.4 Exemplary spectrum with signal and harmonics

Distortion (THD), Signal-to-Noise and Distortion Ratio (SNDR), Dynamic Range (DR), Spurious-Free Dynamic Range (SFDR), Idle Channel Noise (ICN) and the overload level. In general, the overall performance of a delta-sigma converter can be characterized by the dynamic behavior, which can be further divided into spectral, frequency and power domain. The analysis of the power or amplitude spectral density of the output data stream is straightforward, because the proper functionality can be easily proven. Particularly, the delta-sigma converter is mainly analyzed in the frequency domain since it follows the principle of a noise-shaping converter. The calculation or estimation of the output power spectrum is done by the method of discrete Fourier transform, which in turn can be efficiently computed by the Fast Fourier Transform (FFT). In this book, for the estimation of the power spectrum standard MATLAB functions will be employed as shown in later chapters. Beside the shape of the calculated spectrum two major quantitative measures exist for expressing the linearity. First, Total Harmonic Distortion (THD) is the ratio of the sum of the signal power of all harmonic frequencies above the fundamental frequency to the power of the fundamental frequency. Second, Spurious-Free Dynamic Range (SFDR) is defined as the ratio of the signal power to the power of the strongest undesired spectral tone. Both measures are shown in an exemplary signal spectrum in Fig. 2.4. The importance of both measures strongly depends on the targeted application and its sensitivity to harmonics.

The noise and power metrics are the most important measures for delta-sigma converters. These measures are derived from the converter's output spectrum by integration or by proper interpretation. The integrated in-band noise, or the integrated noise of the band of interest (analog bandwidth), is a common converter measure, since it results in the minimum resolvable signal power. It is worth mentioning, the integrated in-band noise contains all in-band power including noise, distortion and other unwanted tones, whereas the desired signal itself is subtracted only. The Signal-to-Noise Ratio (SNR) is the ratio of the signal power to the noise power at the output of the converter, specified for a certain input amplitude. Any occurring

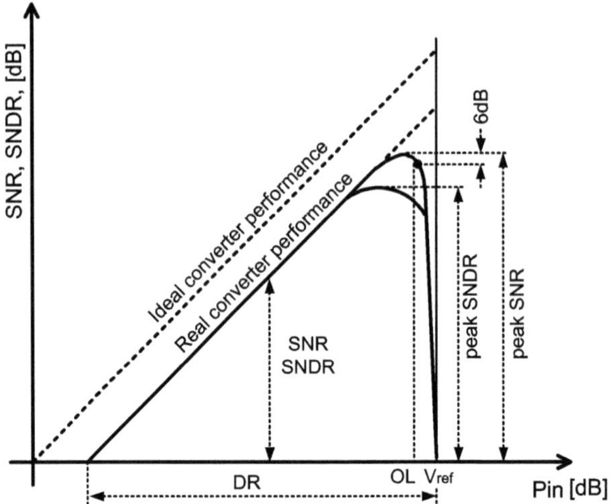

Fig. 2.5 Exemplary performance characteristic

spurs and harmonics are not taken into account. Commonly, the so called peak-SNR is pointed out, because it shows the highest achievable SNR. The Signal-to-Noise and Distortion Ratio (SNDR) is the ratio of the signal power to the noise and all distortion power components. Thus, the corresponding spectra are obtained by applying a signal in the lower frequency region to include at least the second and third harmonic inside the band of interest. The Dynamic Range (DR) is defined as the root mean square (rms) value of the maximum allowable amplitude of the input sinusoidal signal to the root-mean-square value of the smallest detectable input sinusoidal signal. Basically, the provided reference voltage V_{ref} to the converter defines the maximum input amplitude, whereas for delta-sigma modulation even a smaller input amplitude leads to instable operation. This point will be discussed in more detail in the following chapters. In general, the overload level (OL) is defined as the input level, at which the converter still operates correctly. Obviously, this is not a precise definition, but most publications regard the converter to operate correctly until the SNR falls 6 dB below the peak-SNR for higher amplitudes. Usually, for delta-sigma modulators most of these mentioned power measures can be retrieved by generating a plot as illustrated in Fig. 2.5.

The SNR and SNDR is plotted versus the relative input signal power given in decibel (dB). The varied amplitude of the input signal is referred to full-scale, which in turn is equal to the provided reference voltage V_{ref}. This plot is commonly used to prove the performance of a delta-sigma converter.

Beside SNR, another common measure of a converter's accuracy is used which is called Effective Number-of-Bits (ENOB). Assuming a sinusoidal signal applied to an ideal multi-bit quantizer with number of bits B, whereas due to the sampling theorem the sampling frequency is exactly two-times higher than the input frequency and the input amplitude is equal to full-scale level. It can be shown, that the dynamic range of such a Nyquist converter results in a peak-SNR of:

$SNR = 6.02 \cdot B + 1.76$ [dB]. It can be derived that each additional bit results in about 6 dB increase of peak-SNR. Commonly, B is replaced by the Effective Number-of-Bits (ENOB), since in a real converter the physical number of bits do not necessarily equal to the effective number of bits. This leads to following formula for a Nyquist converter:

$$SNR = 6.02 \cdot ENOB + 1.76 \text{ [dB]}. \tag{2.5}$$

While there are many Nyquist-rate converters for high-speed and/or moderate resolution (below 10–12 ENOBs) applications, they all require accurate operations such as comparison, amplification, etc., which sets a tough requirement for the intrinsic precision of the integrated circuits. This fact leads to the principle of over-sampling converters. This means, the sampling frequency is much higher than two-times the maximum input frequency. Employing high Over-Sampling Ratio (OSR) makes the implementation of an anti-aliasing filter easier and more important, the total quantization noise is independent from the sampling frequency. Therefore, the power spectral density of the quantization noise reduces proportionally with increasing sampling frequency, while the integrated part in the band of interest decreases alike. Taking the OSR for an ideal quantizer into account and under the above mentioned assumptions it can be shown that:

$$SNR = 6.02 \cdot ENOB + 10 \cdot \log_{10}(OSR) + 1.76 \text{ [dB]}. \tag{2.6}$$

Therefore, every doubling of the oversampling ratio increases the resolution by 3 dB or about half a bit. This holds as long as the white-noise model for the quantization process is valid. Nevertheless, the approach of increasing the OSR is not really profitable, because the improvement of resolution is poor compared to the required increase in power drain for achieving faster conversion rates in real circuits. However, the accuracy requirements on the analog components are relaxed compared to those associated with Nyquist-rate converters. The cost paid for high accuracy thus includes faster operation, but this is getting cheaper as integrated circuits technology advances.

The definition of a Figure Of Merit (FOM) is often taken into consideration to be able to compare different converter solutions with each other. The most commonly used FOM takes the relation of power consumption over converter resolution and band of interest. In the following chapters, the following FOM will be used:

$$FOM = \frac{P}{2 \cdot BW \cdot 2^{ENOB}}. \tag{2.7}$$

In this FOM, a factor of two in power (P), a factor of two in band of interest or bandwidth (BW) and one effective number of bit (ENOB) are all considered to be equivalent. This FOM is typically expressed in pJ/conversion-step.

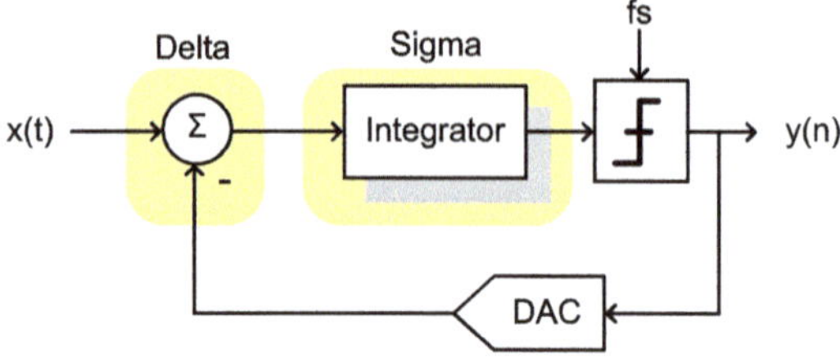

Fig. 2.6 A simple delta-sigma modulator

2.3 Delta-Sigma Converter

In a straightforward manner, A/D-converters might be considered to be in open-loop configuration such as flash-converters, successive-approximation-converters (SAR) or pipeline-converters. In that case, an output signal is steered by an input signal and affected by quantization noise. The resulting system seems to be quite simple and easy to understand, but all undesired effects like distortion and noise contribution will be added to the signal path and deteriorates the overall performance. A general approach to overcome this drawback is the introduction of a feedback path to achieve a closed control-loop. The two main types of oversampling converters in feedback configuration are the delta modulator and the delta-sigma modulator. (They are called modulator or converter loop as well, since these converters contain several filter stages to perform the actual analog to digital conversion.) The focus of this book lays in the field of delta-sigma converter, hence the principle of delta modulation can be found in [54]. A very simple example for a delta-sigma modulator is depicted in Fig. 2.6. It is a feedback loop, containing an internal low-resolution A/D- and D/A-converter, as well as a loop filter, in this example an integrator only. It is a nonlinear system due to the quantization effect as well as a dynamic system due to the memory in the integrator. Both facts make the analysis to a difficult mathematical task.

Simple qualitative understanding of its operation can be gained by using the linearized noise-model as already discussed in this chapter, which consists of a unity-gain buffer an additive quantization noise e. Assuming perfect operation of the D/A-converter as well as a reference voltage of $V_{ref} = 1$ V and a sampling rate $f_S = 1$ Hz, the discrete-time linear system of Fig. 2.7 results. Two different transfer functions can be obtained for the disturbing noise as well as the desired signal, defining a Signal Transfer Function (STF) and a Noise Transfer Function (NTF). An optimal configuration for the STF and NTF is the complete extinction of the noise inside the desired signal bandwidth, while herein the signal itself is not affected. Outside the band of interest the NTF gain can be high, while the STF can be low instead.

Analysis of Fig. 2.7 gives following transfer functions:

$$STF = \frac{Y}{X} = z^{-1} \quad \text{for } E = 0, \tag{2.8}$$

Fig. 2.7 A z-domain model
of a simple modulator

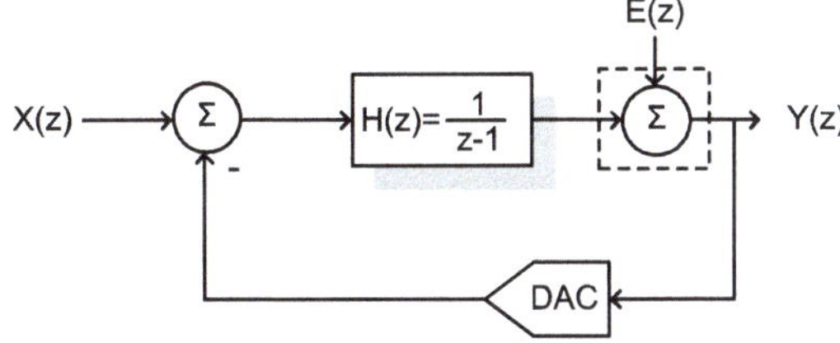

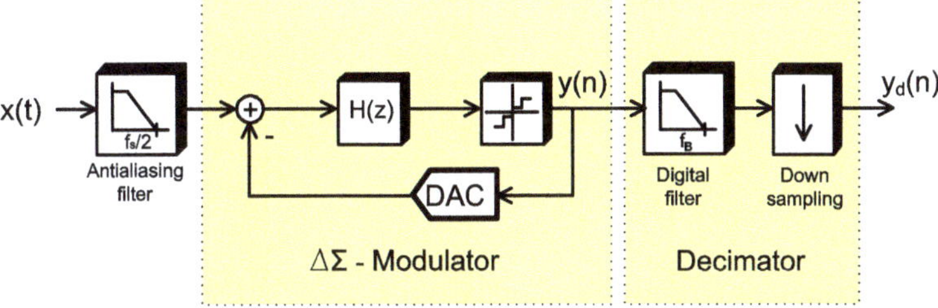

Fig. 2.8 Block diagram of a delta-sigma converter

$$NTF = \frac{Y}{E} = \frac{1}{1 + \frac{1}{z-1}} \quad \text{for } X = 0. \tag{2.9}$$

Its STF and NTF yield:

$$Y(z) = X(z) \cdot STF(z) + E(z) \cdot NTF(z). \tag{2.10}$$

In the above calculation the DAC is assumed to be ideal, while $X(z)$, $Y(z)$ and $E(z)$ are the z-domain representations of the input and output signal and the quantization error. In general, if the loopfilter has a high gain in the signal band, the in-band quantization noise is strongly attenuated, a process commonly called noise shaping. That is the reason why for the simple delta-sigma modulator an integrator as a loopfilter was chosen. Note, the loopfilter is in the forward path of the loop, hence its nonidealities are strongly suppressed by the loop gain. Any nonlinearity of the coarse A/D-converter is simply combined with the quantization error e, and is thus suppressed in-band along with e. Nonlinear distortion in D/A-converter, however, remains after subtraction from the input signal and affects the output signal without any shaping. Hence, the D/A-converter represents a major limitation on the attainable performance.

It is worth mentioning, that the delta-sigma modulator alone does not yet represent an A/D-converter, since the output data are given at high data-rates and small bit-widths. Figure 2.8 shows all relevant components comprising a whole A/D-conversion path.

The signal enters into the anti-aliasing filter eliminating spectral components above half the sampling frequency to avoid any down-folding of high frequency tones into the band of interest, as expressed in the Nyquist theorem (2.1). Thanks to the oversampling ratio, the lowpass filter needs not a sharp transition between the passband and the stopband, which helps to reduce the filter complexity. The band limited signal is applied to the delta-sigma modulator to perform the actual A/D-conversion by means of sampling and quantizing. The quantizer is of low resolution, typically 1 to 5-bits are used for competitive product designs to keep the occupied silicon area small. The internal D/A-converter in the feedback path is commonly implemented with the same low resolution as the internal quantizer. The delta-sigma modulator provides a noise-shaped signal showing small in-band quantization noise. The output signal runs at high oversampled data-rates and due to the principle of noise-shaping the out of band noise gets strongly amplified. Additional lowpass filtering is needed in the digital domain. A down-sampling process is required to get a digitized signal at Nyquist data-rate with an output bit-width appropriate to the overall A/D-converter resolution. Commonly, the lowpass filtering along with the down-sampling is done together in a digital decimation filter chain.

Note, the quantizer receives a clock signal and provides a uniformly sampled sequence of sample points. The sampling process turns the modulator into a synchronous system due to the nature of a regular sampling pattern. At this point inside the modulator the transformation takes place from a continuous-time domain into a discrete-time domain. Thus, this kind of modulator can be categorised into a *synchronous* delta-sigma modulator.

In general, the principle of delta-sigma modulation can be employed for actual D/A-converters as well. This approach allows a much higher resolution as compared to Nyquist-rate D/A-converters. By operating a fully digital delta-sigma modulator loop at an oversampled clock-rate, a data stream with commonly 14-bit to 20-bit word-length my be changed into a single-bit digital signal such that the baseband spectrum is preserved. The large amount of truncation noise generated in the digital loop is shaped in order to make the in-band noise negligible [54]. This book deals with A/D-conversion and hence the focus is put on analog delta-sigma modulators.

2.3.1 Single-Loop Modulator

As already introduced in Fig. 2.6 the simplest analog delta-sigma modulator is a first-order loop containing an integrator, a single-bit quantizer and a single bit D/A converter. A closer look on the NTF unveils the following simplification in the z-domain:

$$NTF(z) = \left(1 - z^{-1}\right). \tag{2.11}$$

Replacing z by $e^{j2\pi fT}$ to get into the frequency domain, the Power Spectral Density (PSD) of the output noise is found to be

$$S_q(f) = \left(2\sin(\pi fT)\right)^2 \cdot S_e(f). \tag{2.12}$$

Here, $T = 1/f_s$ is the sampling period, and $S_e(f)$ is the single-sided PSD of the quantization noise of the internal A/D converter. As already discussed in (2.4), the

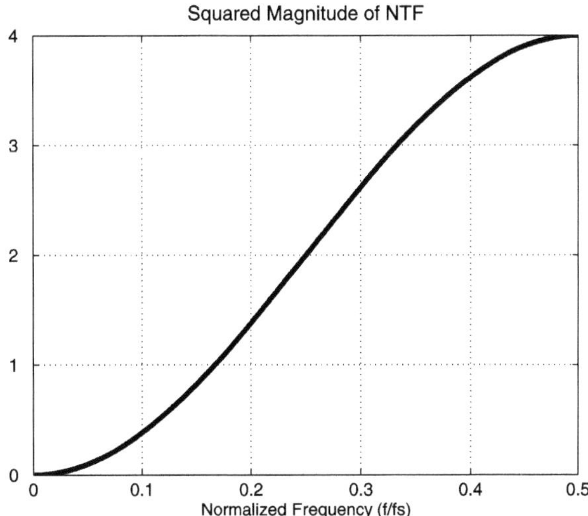

Fig. 2.9 NTF for a simple delta-sigma modulator

white noise model of a quantizer with a step size of Δ approximates the error with white noise of mean square value $e^2_{rms} = \frac{\Delta^2}{12}$ and thus

$$S_e(f) = \frac{\Delta^2}{6 f_S}. \tag{2.13}$$

The NTF can be seen as a filtering function with highpass characteristic. It suppresses the quantization error at frequencies around 0, but the NTF also enhances the quantization error at higher frequencies around $\frac{f_S}{2}$, as illustrated in Fig. 2.9. Taking the OSR into account, that means integrating $S_q(f)$ between 0 and f_B gives the in-band noise power. Assuming $OSR \gg 1$, the In-Band Quantization Noise (IBQN) becomes:

$$IBQN = \int_0^{f_B} \frac{\Delta^2}{6 f_S} \left(2 \sin(\pi f T)\right)^2 df \approx \frac{\Delta^2}{12} \cdot \frac{\pi^2}{3} \frac{1}{OSR^3}. \tag{2.14}$$

As expected, the in-band noise decreases with increasing OSR. However, this decrease is relatively slow, since doubling the OSR reduces the noise only by 9 dB, and hence improves the ENOB by only about 1.5 bits.

An obvious way to increase the resolution is to use a higher-order loopfilter. By adding another integrator and feedback path to the first-order modulator, Fig. 2.10 results.

Employing the linear noise model for the quantizer to get a linearized representation of the second-order delta-sigma modulator leads to following equation:

$$Y(z) = z^{-1} \cdot X(z) + \left(1 - z^{-1}\right)^2 \cdot E(z). \tag{2.15}$$

This equation shows that the NTF is now $(1 - z^{-1})^2$, which is equal to a noise shaping function of $(2 \sin(\pi f T))^4$. Assuming $OSR \gg 1$, a good approximation for

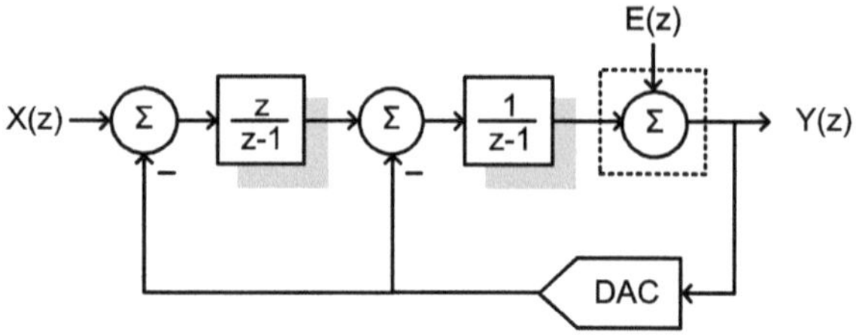

Fig. 2.10 A second-order delta-sigma modulator

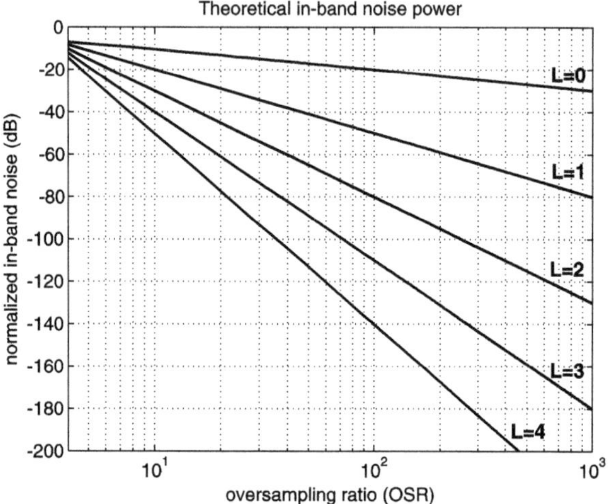

Fig. 2.11 In-band noise for Lth-order modulators

IBQN becomes:

$$IBQN = \int_0^{f_B} \frac{\Delta^2}{6 f_S} \left(2\sin(\pi f T)\right)^4 df \approx \frac{\Delta^2}{12} \cdot \frac{\pi^4}{5} \frac{1}{OSR^5}. \qquad (2.16)$$

Hence, doubling of OSR results in about 2.5 bits of additional dynamic-range. This is much more as compared to that of the first-order modulator. In principle, by adding more integrators and feedback branches to the loop, even higher-order NTFs can be obtained. For an Lth-order loop filter resulting in $NTF(z) = (1 - z^{-1})^L$, the in-band noise power is approximately ($OSR \gg 1$):

$$IBQN \approx \frac{\Delta^2}{12} \cdot \frac{\pi^{2L}}{2L+1} \cdot \frac{1}{OSR^{2L+1}}. \qquad (2.17)$$

Hence, doubling the OSR gives about $L + 0.5$ additional bits to the resolution. The theoretical in-band noise as a function of OSR is plotted in Fig. 2.11 for different Lth-order delta-sigma modulators. In this plot 0 dB corresponds to a quantization noise power of $\frac{\Delta^2}{12}$.

It is worth mentioning that stability issues have been ignored so far. Note, for high-order loops, stability considerations reduce the achievable resolution to a lower value than that given by the above equations. This topic will be discussed in detail in the next chapters. One possible solution to avoid stability issues is to go for multi-bit quantizers. So far, only single-bit quantizers have been considered. Any occurring nonidealities in the feedback D/A-converter result in comparable nonidealities for the overall conversion. This is because the output signal of the D/A-converter is forced by the feedback loop to follow the input signal $x(t)$ very accurately. Hence, any nonidealities in the D/A-converter result in a distorted signal at the input of the loopfilter and deteriorates the overall performance. It was this fact which forced early delta-sigma designs to use single-bit internal quantizers and D/A-converters. However, single-bit quantizers have an undefined gain factor, since they are essentially comparators only. Keeping such modulator loops stable results in a reduction of the allowable input signal swing. This is because loops containing single-bit quantizers must remain stable over a wide range of loop gains. In contrast, for a multi-bit quantizer, the loop is inherently more stable since the quantizer gain is well-defined. In fact, linear analysis can be used to design the modulator so that stability is guaranteed. Furthermore, since the quantization noise decreases by 6 dB for each bit added to the quantizer, multi-bit modulators may have very high ENOB even at low OSR values. This fact is a strong motivation to solve the problem of D/A nonlinearity in the use of multi-bit quantization. While brute-force techniques have been used earlier, such as element trimming, the more currently techniques use additional digital circuitry to manipulate the elements of D/A-converters in such a manner so that the in-band portion of the error gets reduced. These techniques are conceptionally very similar to the noise shaping used in delta-sigma modulators, and are described in the open literature with the term *'mismatch shaping'*. A common approach for mismatch shaping is the so called Dynamic Element Matching (DEM). As with noise shaping, the effectiveness of mismatch shaping increases with higher OSR.

2.3.2 Cascaded-Loop Modulator

Another way of achieving high resolution, which eases the stability problems associated with high-order modulators, is the approach of cascading of several modulator stages. These types are called Multi-stAge noise-SHaping (MASH) or cascaded-loop modulators. The basic concept is shown in Fig. 2.12.

The output signal of the first delta-sigma stage is given by

$$Y_1(z) = STF_1(z) \cdot X(z) + NTF_1 \cdot E_1(z). \tag{2.18}$$

STF_1 and NTF_1 are the signal and noise transfer functions, respectively, of the first delta-sigma stage. The quantization error $E_1(z)$ is found in analog form by subtracting the input to its internal quantizer from its output. $E_1(z)$ is fed to another delta-sigma loop forming the second stage of the cascaded-loop modulator, and converted

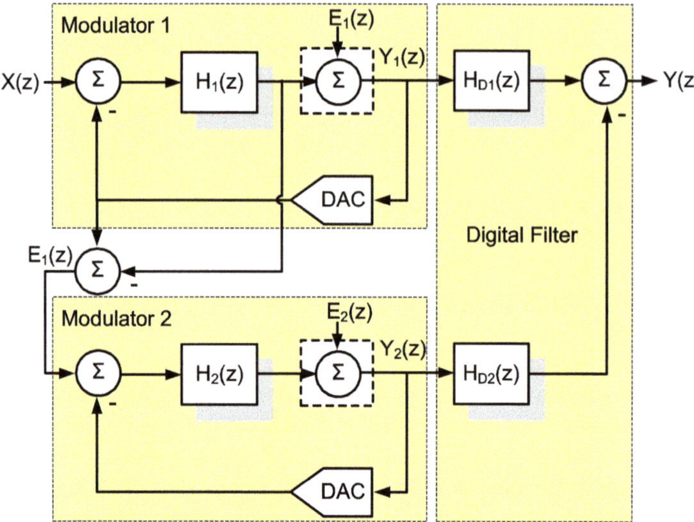

Fig. 2.12 A cascaded-loop modulator

into digital form. Hence, the output signal of the second stage in is given by

$$Y_2(z) = STF_2(z) \cdot E_1(z) + NTF_2 \cdot E_2(z). \qquad (2.19)$$

STF_2 and NTF_2 are the signal and noise transfer functions, respectively, of the second delta-sigma stage. The digital filter stages $H_{D1}(z)$ and $H_{D2}(z)$ at the outputs of the two modulator loops are designed such that in the overall output $Y(z)$ the quantization error $E_1(z)$ is canceled. By above given equations $Y_1(z)$ and $Y_2(z)$, this is achieved if following condition holds:

$$H_{D1} \cdot NTF_1 - H_{D2} \cdot STF_2 = 0. \qquad (2.20)$$

A practical choice for $H_{D1}(z)$ and $H_{D2}(z)$ which satisfies the condition above is $H_{D1}(z) = k \cdot STF_2$ and $H_{D2}(z) = k \cdot NTF_1$, where k is a constant to give unity signal gain. Since STF_2 is often just a delay (e.g. z^{-1}), $H_{D1}(z)$ can be easily realized. The overall output of the MASH is then given by:

$$Y(z) = H_{D1}(z) \cdot Y_1(z) - H_{D2}(z) \cdot Y_2(z), \qquad (2.21)$$

$$Y(z) = k \cdot STF_1 \cdot STF_2 \cdot X(z) - k \cdot NTF_1 \cdot NTF_2 \cdot E_2(z). \qquad (2.22)$$

In the open literature very often a second-order loop is chosen for both stages of the MASH modulator. Taking this as an example, the signal transfer functions of the first and second loop my be given by $STF_1 = STF_2 = z^{-1}$. The noise transfer functions are given by $NTF_1 = NTF_2 = (1 - z^{-1})^2$. Choosing $k = 1$, the output is then

$$Y(z) = z^{-2} \cdot X(z) + \left(1 - z^{-1}\right)^4 \cdot E_2(z). \qquad (2.23)$$

That means, the performance of the noise-shaping behavior is essentially that of a fourth-order single-loop converter, but the stability behavior is that of a second-order modulator. This fact has put high importance for designing A/D-converters employing the MASH topology. On the other hand, there is a limitation in using that approach. If the condition $H_{D1} \cdot NTF_1 - H_{D2} \cdot STF_2 = 0$ is not exactly satisfied due to imperfections in the realization of the analog transfer functions, then $E_1(z)$ will appear at the output $Y(z)$ multiplied by $k(STF_2 \cdot NTF_{1a} - NTF_1 \cdot STF_{2a})$. NTF_{1a} and STF_{2a} describe the actual value of the analog transfer functions. These unmatched transfer functions will result in a serious deterioration of the noise performance of the A/D-converter. This unwanted effect is called '*noise leakage*' in cascaded modulators. Achieving very accurate analog transfer functions require high power drain for the active stages and consume more chip area for good device matching, which makes MASH topologies to be traded off against the beneficial stability behavior.

2.3.3 Discrete-Time vs. Continuous-Time Loopfilter

In the previous subchapters, the loop transfer function $H(z)$ has been written in the discrete-time domain. The majority of delta-sigma modulators in the literature are implemented as discrete-time circuits such as Switched-Capacitor (SC) circuits. It is possible to build the loopfilter as a continuous circuit $H(s)$, which becomes more and more important at high frequencies. There are several reasons why a loopfilter is chosen to be in *discrete-time* domain or in *continuous-time* domain. Perhaps the major reason for the prevalence of discrete-time based modulators is that there is a natural allegory between the mathematics of the system and its circuit level implementation. Fundamentally, delta-sigma modulators rely on having at least one integrator inside the loop which requires to construct a circuit to result in $\frac{1}{z-1}$ integrating function. This function can be easily transformed into a SC-integrator. One general issue with A/D-converters is *aliasing* of high frequency tones into the band of interest, as already discussed in Nyquist's theorem (2.1). A discrete-time loopfilter samples the signal right at the input of a modulator stage. Thus, a separate filter is required at the input to attenuate aliases sufficiently. On the other hand, modulators comprising continuous filters sample the signal at the input of the quantizer, in other words, at the output of the loopfilter. This fact can be employed to include the anti-aliasing functionality inside the existing continuous-time loopfilter.

Usually, continuous modulators show a higher sensitivity to *clock jitter*, since any displacement in time of the signal in the feedback path causes an error proportional to the amount of displacement in time. In switched-capacitor circuits the signal needs to be settled at the end of the clock-cycles since the transfer of charges among several capacitors represents the desired signal computation. The shape of the voltage curve within the clock-cycle is not relevant as long as the settling transient reaches the desired value at the end of the clock-cycle. Thus, any occurring clock jitter does not deteriorate the performance inside the switched capacitor filter. However, the signal is sampled at the input and clock jitter might cause a performance loss, but thanks to the OSR the loss is much smaller as compared to a continuous-time loop filter implementation [24, 54].

Basically, in a SC circuit large glitches appear on Operational Amplifier (OpAmp) *virtual ground* nodes and output nodes due to switching transients. Maintaining sufficient settling accuracy of all circuit waveforms, OpAmps with high bandwidth and high slew-rate are required which causes high power drain. By contrast, waveforms vary much more smoothly in a continuous-time loopfilter which can be employed to reduce the power drain or to increase the OSR for achieving higher resolution [56].

In practice, the quantizer latch and the feedback D/A-converter cannot switch simultaneously. That means, a certain delay between the quantizer clock and D/A-converter waveform exists, which is named in the literature as Excess Loop Delay (ELD). The delay causes a shift of the D/A feedback signal which affects the entire continuous-time loopfilter characteristic. Furthermore, the implemented feedback D/A-converter also exhibits slew-rate limitations as well as different rise and fall times of the pulses. Both nonidealities tend to degrade the performance, unless these errors are made sufficiently small. Especially in continuous-time modulators, the impact of both errors causes intersymbol interference due to the fact that the resulting errors depend on the output sequence of the modulator. This effect causes additional noise and unwanted tones within the spectrum that fold into the band of interest.

The following bulleted list gives a summary for the main advantages for both implementations:

- Continuous-time loopfilter
  - Relaxed OpAmp speed requirements
  - Higher sampling frequency possible
  - Lower power drain possible
  - Implicit anti-aliasing filtering
  - Reduced switching noise and smaller glitches on supply rails
  - Lower simulation time on SPICE-level
- Discrete-time loopfilter
  - Allegory between mathematics and CMOS circuits
  - Accurately defined filter coefficients
  - Low sensitivity to clock-jitter
  - Low sensitivity to DAC waveform
  - Low sensitivity to Excess Loop Delay (ELD)
  - Capacitive loads only

2.3.4 Circuit Nonidealities for Discrete-Time Loopfilters

The performance of a practical implementation of a $\Delta\Sigma$-converter can be significantly worse than the values calculated with the previously derived formulas. That

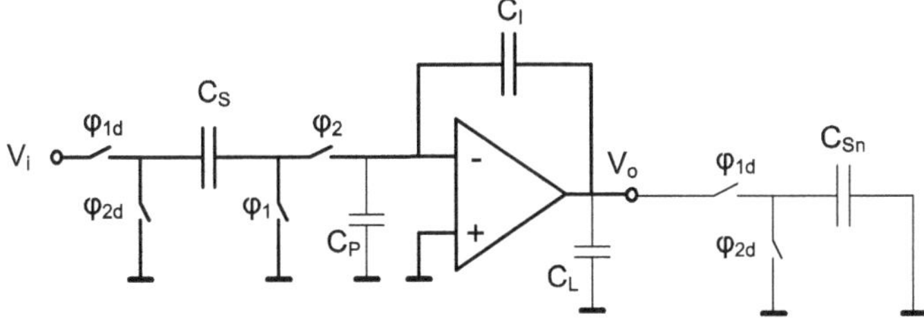

Fig. 2.13 A model of a switched-capacitor integrator

is because circuits show nonidealities influencing the desired behavior. Ensuring a proper design requires a know-how about the influence of all nonidealities on the performance of the $\Delta\Sigma$-converter. The basic building block of a loopfilter is an integrator. A conventional switched-capacitor integrator is depicted in Fig. 2.13. Several important nonidealities of the switched-capacitor integrator are discussed, whereby the derivation of the models and equations are not discussed in this book, but a detailed description can be found in [76] and [77].

The single-ended model of the switched-capacitor integrator is shown in Fig. 2.13. Any possible implementation of a feedback D/A-converter is not shown here to keep the presentiveness. V_i and V_o represents the input and output voltage, C_S and C_I the sampling and integration capacitors, respectively. In order to obtain an accurate model, it is very important to include the parasitic capacitances C_P and C_L associated with the input and output capacitances of the Operational Transconductance Amplifier (OTA). C_L also includes the bottom plate parasitic of C_I. C_{Sn} is included to model the sampling operation of the next stage, whereby the top plate of C_{Sn} is connected to a perfect *virtual ground* during both clock phases.

The following list gives an overview about the main circuit nonidealities for a switched-capacitor integrator:

- Finite OTA DC-gain
- Finite dominant closed-loop pole of the OTA
- Finite switch-resistance
- Limited slew-rate

First, the effect of finite gain of the OTA is discussed briefly. The OTA is represented by a voltage controlled voltage source with gain A. It is worth mentioning, that in real OTAs the gain depends strongly on the frequency such as for a lowpass filter with a certain gain. By applying the methods described in [76] and [77], the output voltage of the integrator at the end of the sampling phase can be calculated as

$$V_o(z) = \frac{C_S}{C_I} \cdot \frac{\rho_2 \cdot z^{-1} V_i(z)}{1 - \frac{\rho_2}{\rho_1} z^{-1}} \tag{2.24}$$

where ρ_1 and ρ_2 are the closed-loop static errors and λ_1 and λ_2 are the capacitive feedback factors during sampling and integration phase, respectively. They are given by following equations:

$$\rho_1 = \frac{A_0\lambda_1}{1 + A_0\lambda_1}, \qquad \rho_2 = \frac{A_0\lambda_2}{1 + A_0\lambda_2}, \tag{2.25}$$

$$\lambda_1 = \frac{C_I}{C_P + C_I}, \qquad \lambda_2 = \frac{C_I}{C_S + C_P + C_I}. \tag{2.26}$$

The finite OTA DC-gain introduces two errors in the transfer function of the integrator. The gain error reduces the gain of the integrator by ρ_2 and the pole error moves the pole from DC ($z = 1$) to $z = \frac{\rho_2}{\rho_1}$. Both errors depend on the product of the OTA gain and the capacitive feedback factors. This model can be used to determine the required OTA DC-gain. In a practical implementation of an OTA, the gain is not the same for all values of the output voltage. Instead, it decreases as the output voltage increases. This is due to the reduction of the output resistance as the drain-source voltage of the output transistors decreases. The nonlinear gain of the OTA can be modeled by a truncated Taylor expansion as

$$A(v) = A_0 \cdot \left(a_1 v + a_2 v^2 + a_3 v^3\right) \tag{2.27}$$

where v is the output voltage of the OTA. Note that a_2 and a_3 are negative values since the gain decreases as the output swing increases. This polynomial can be used for behavioral simulations to derive the required linearity of the OTA. As an example for the polynomial coefficients, values of state of the art designs are given, such as $a_1 = 1$, $a_2 = -0.003$ and $a_3 = -0.04$. These values were derived from a fully-differential two-stage Miller-OpAmp designed in a 65 nm technology.

In a practical implementation, the poles of an OTA limit the settling performance. Thus, all voltages show a transient response during the sampling and integration phase. The settling transient depends on the position of the poles. Therefore, the OTA is modeled with a transconductance g_m and a finite output conductance g_o to study the influence of the dominant closed-loop pole of the amplifier. The dominant closed-loop pole p_{cl2} during the integration phase can be calculated as

$$p_{cl2} = \frac{g_m}{C_{eq,i}} \tag{2.28}$$

where $C_{eq,i}$ represent the effective capacitive load of the OTA during the integration phase and can be calculated as

$$C_{eq,i} = C_S + C_P + \frac{(C_S + C_P + C_I)C_L}{C_I}. \tag{2.29}$$

This model can be used for behavioral simulations to derive the required speed of the OTA.

Up to this point, all switches have been assumed to have an ideal zero resistance when they are supposed to be closed. In CMOS technologies, they are implemented with nMOS and/or pMOS transistors exhibiting several non-ideal effects such as a non-zero resistance, clock feedthrough, charge injection and the variation of the resistance with the applied voltage. During both clock phases two switches are set in series and the resistances are lumped into one resistance, R_{sw1} and R_{sw2} for the sampling and integrating phase, respectively. Compared to the equation for the settling error due to limited OTA speed, which does not include the switch resistance, the dominant closed-loop pole is degraded to

$$p_{cl2,R} = \frac{p_{cl2}}{1 + p_{cl2} \cdot R_{sw2} C_S}.$$

(2.30)

This clearly shows the influence of the switch-resistance during the integration phase on the degradation of the settling behavior of the integrator. The second important problem related to the switches is the variation of the switch-resistance with the input signal. When a switch is in the on-phase, it can be assumed that the transistor is in the linear operation region. The channel-resistance between source and drain of a nMOS transistor is given by [72] and [71]:

$$R_{on} = \frac{1}{K'(\frac{W}{L}) \cdot (v_G - V_{TH} - \frac{v_S + v_D}{2})}.$$

(2.31)

K' and V_{TH} are technology constants, the design parameter W (width) and L (length) define the size of the transistor and v_S, v_D and v_G are the node voltages at the source, drain and gate terminal, respectively. This equation shows two important effects. First, a reduction of the supply voltage immediately increases the switch-resistance since the possible overdrive voltage ($v_G - V_{TH}$) of the transistor decreases. Second, the switch-resistance is dependent on the source and drain voltages. When looking at the input switch of Fig. 2.13, it is obvious that the on-resistance depends directly on the input signal of the integrator. This will generate harmonic distortion. Both effects can be reduced by proper sizing and/or by employing transmission gates with nMOS and pMOS transistors in parallel.

Another important nonideal effect in switched-capacitor integrators is the slewing of the OTA. The slewing performance of the integrator is dependent on the occurring node voltages. Slewing is most likely to occur at the beginning of the integration phase. The voltage sampled on the sampling capacitor C_S is switched between the OTA's feedback signal (the voltage across C_I) and the input terminal of the OTA, causing a large voltage spike which most probably drives the OTA into slewing operation (see Fig. 2.13). The OTA drives the maximum output current to accommodate charge redistribution between C_S and C_I until the voltage difference between both input terminals of the OTA gets sufficiently small for recovering into normal operation of the OTA. This effect has to be taken into account for designing the required OTA performance accordingly to guarantee proper settling to not compromise the targeted loop filter transfer function.

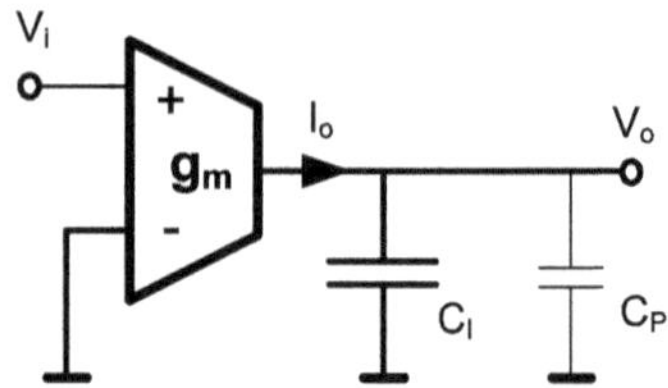

Fig. 2.14 A model of a $g_m C$-integrator

2.3.5 Circuit Nonidealities for Continuous-Time Loopfilters

The loopfilter transfer function is the main performance determining part in a delta-sigma modulator, because it defines the noise-transfer function and therewith the quantization noise-shaping behavior. Basically, a loopfilter consists of several first-order filters, which are commonly arranged in a feedback or forward architecture. Taking continuous-time modulators into consideration, the single filters are realized using either RC-integrator or $g_m C$-integrator or even LC-resonators for bandpass noise shaping. A $g_m C$-integrator is based on a transconductance amplifier and a capacitor as shown in Fig. 2.14. The input voltage V_i is fed through a transconductor g_m producing a current $I_o = g_m \cdot V_i$ to drive the capacitance C_I. The ideal transfer function of a $g_m C$-integrator yields:

$$H(s) = \frac{g_m}{sC_I}. \tag{2.32}$$

For high-frequency filters the g_m-stage can be reduced to its simplest possible configuration leading to a differential pair, to which some linearization techniques can be applied [71]. The transconductance of such a g_m-stage directly depends on the biasing current. If the differential pair MOSTs work in strong inversion, the g_m is proportional to the square root of the current. Furthermore, operating the MOSTs in weak inversion, the g_m is directly proportional to the bias current, which is the same as for bipolar transistors. Thus, filter tuning can be realized relatively easy. Due to its simplicity, these circuits are able to operate to very high frequencies, which is one of the major advantages. However, maintaining full-swing at the output to optimize the dynamic-range results in increased distortion components, due to the nonlinearities in the g_m-stage and basically to the nature of an open-loop structure. A commonly used method to increase the linearity is *source degeneration*, where a resistor is placed between both source nodes of the differential input pair, as shown in Fig. 2.15. The effective transconductance of such a g_m-stage can be calculated as:

$$g_{m,eff} = \frac{g_m}{1 + g_m \cdot R_{deg}} \tag{2.33}$$

Note that increasing of R_{deg} leads finally to an effective transconductance which is proportional to $1/R_{deg}$. This means, the effective transconductance is determined by a passive resistor only, which is in most cases sufficiently linear. However, that technique lowers the DC-gain of the OTA and further on, the resistive degeneration

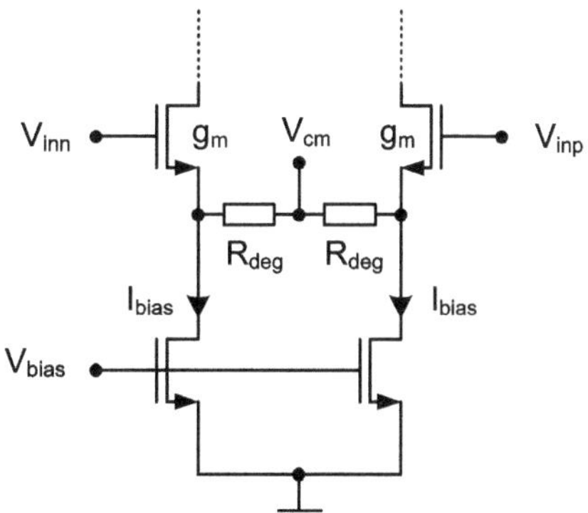

Fig. 2.15 Resistive source degeneration of a g_m-stage

increases the contribution of transistor noise in the total input noise power of the integrator. Thus, there is a trade-off between the suppression of thermal noise and harmonic distortion components. In [50] following relation has been derived between the bias-current I_{bias} and the transconductance g_m of the differential-pair[1] as well as the degeneration resistor R_{deg} for estimating the SNDR of a g_mC-integrator:

$$SNDR_{gmC} \sim I_{bias} \cdot g_m \cdot R_{deg}. \tag{2.34}$$

Hence, the $SNDR_{gmC}$ is proportional to the bias-current and the degeneration value.

As a result this structure is only conditionally suitable for low-voltage design, since a reduction of the signal-swing requires a reduction of the thermal noise to keep the same dynamic-range. Another drawback is the sensitivity to the parasitic capacitance C_p (see Fig. 2.14), which directly alters the integrator time constant. Generally speaking, for linearity requirements up to about $THD = -50 \dots -60\,dB$, g_mC-filters are favorable, if low power is a major interest [35]. In case of demanding a higher linearity, another circuit such as an active RC-integrator is the preferred approach.

High resolution delta-sigma converters require a highly linear integrator such as an active RC-integrator comprising an OpAmp in local-feedback configuration. This approach is preferable to open-loop structures, such as g_mC-integrators, in case of low supply voltage and tight distortion specifications. A conventional RC-integrator is depicted in Fig. 2.16. For simplicity, a single ended active integrator is considered only. In order to obtain an accurate model, it is very important to include the parasitic capacitances C_P and C_L associated with the input and output capacitances of the OpAmp. C_P represents the sum of all parasitics at the *virtual ground* such as input capacitance of the OpAmp, the wiring capacitances of the connected metal wires as well as parasitics associated with the devices R_I and C_I. C_L also includes the bottom plate parasitic of C_I. R_L is included to model the input resistance of the next stage.

---

[1]The product of the transconductance and the degeneration resistance, $g_m \cdot R_{deg}$, is called the *degeneration value*.

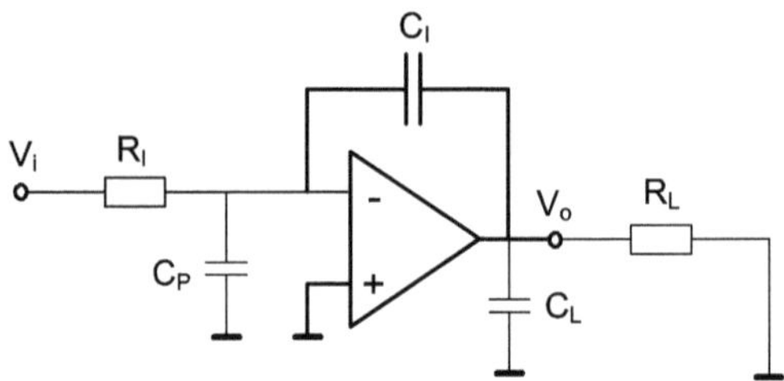

Fig. 2.16 A model of a RC-integrator

Several important nonidealities of the continuous-time RC-integrator are discussed, whereby the derivation of the models and equations are not discussed in this book, but a detailed description can be found in [62] and [61]. The following list gives an overview about the main circuit nonidealities for a RC-integrator:

- Finite DC-gain of the OpAmp
- Finite gain-bandwidth-product (GBW) of the OpAmp
- Parasitic capacitance at virtual ground node of the OpAmp
- Presence of a second non-dominant pole originating from the OpAmp or parasitic input capacitance C_P

Consequently, the parasitic input capacitance has to be kept as small as possible, while the non-dominant pole of the OpAmp has to be shifted towards higher frequencies in order to maintain sufficiently small phase-shift to preserve a stable operation of the $\Delta\Sigma$-modulator. Note, in a RC-integrator a positive zero appears that creates an unwanted phase-shift of $-90°$ [50]:

$$\omega_z = \frac{g_m}{C_I} \tag{2.35}$$

whereas g_m is the transconductance of the OpAmp. This zero can be made negative by placing a small resistor R_x in series with the integrating capacitor. The resistor creates a negative zero[2] instead of a positive zero if following condition is met:

$$\frac{1}{g_m} \ll R_x \ll \frac{g_m}{g_o} \cdot R_I \tag{2.36}$$

whereas g_o stands for the finite output conductance of the OpAmp. A RC-integrator is used in closed-loop configuration resulting in a *virtual ground* at the input of the OpAmp as long as the loop-gain is much higher than unity for all applied frequencies. The distortion components can be kept small for sufficiently high GBW's, however, if the unity-gain frequency is lowered close to the sampling frequency f_s or even below f_s, a steep roll-off of the linear performance is noticed. In [50] it is shown that the SNDR of a RC-integrator is proportional to the bias-current of the differential pair and to the square of $g_m \cdot R_I$:

$$SNDR_{RC} \sim I_{bias} \cdot (g_m \cdot R_I)^2. \tag{2.37}$$

---

[2]Note, a negative zero shifts the phase towards positive values causing less delay.

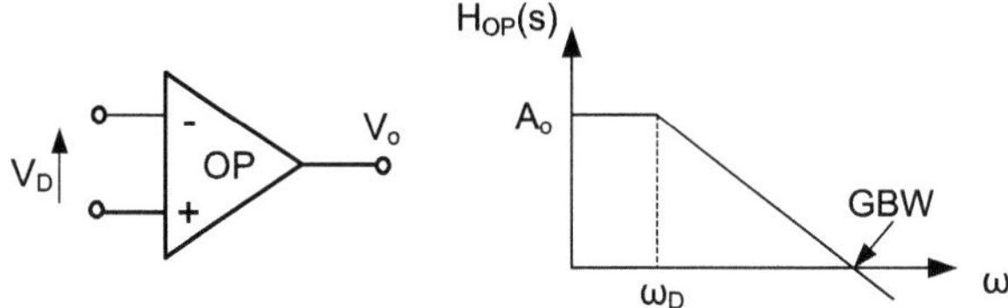

Fig. 2.17   A model of a real OpAmp

As a consequence, for the same SNDR, $g_m C$-integrators require more power than their RC counterparts $\rightarrow$ compare (2.34) and (2.37). Hence, the RC-integrator is a more efficient choice for applications that demand a high conversion accuracy. Nevertheless, the suppression of harmonic distortion components requires sufficiently high loop-gain therewith the input of the OpAmp can be considered as *virtual ground*. Thus, the constraint $GBW > f_s$ must be satisfied leading to an increase of power consumption for very high sampling frequencies. In that case, a $g_m C$-integrator might be a better choice in terms of power drain, as long as a *large sampling frequency* along with rather *moderate accuracy* is demanded.[3]

This book will consider converters with high accuracy only, therefore, the focus will be on loopfilters employing RC-integrators. In order to obtain the right specifications for the circuit design, the main nonidealities of the RC-integrator needs to be modeled and their impact needs to be simulated at system level. A real OpAmp shows finite DC-gain A_0 and finite Gain-BandWidth Product (GBW), which can be modeled in a high-level description for deriving a proper specification.

A simple model for a real OpAmp is depicted in Fig. 2.17, whereas $H_{OP}(s) = \frac{V_O(s)}{V_D(s)}$ is the transfer function of the OpAmp in the s-domain ([72] and [71]). ω_D stands for the dominant pole of the OpAmp. Following equations are used for a high-level description:

$$GBW = A_0 \cdot f_D = A_0 \cdot \frac{\omega_D}{2\pi}, \qquad H_{OP}(s) = \frac{A_0 \cdot \omega_D}{s + \omega_D}. \qquad (2.38)$$

In nanoscale technologies the gain at DC becomes small, for example, in a 65 nm node $A_0 \approx 40$ dB. As consequence, the dominant pole ω_D results in the range of a few MHz. Therefore, ω_D needs to be taken into account since it can be placed within the band of interest. The transfer function of an ideal RC-integrator can be expressed as $H(s) = \frac{1}{s \cdot RC}$. It can be demonstrated that for sufficiently high DC-gain A_0 of the OpAmp, the real RC-integrator can be modeled as a cascade of an ideal integrator and a first-order lowpass filter. The effect of the nonidealities can then be lumped

---

[3]Obviously, the exact numbers for a "large" sampling frequency and a "moderate" accuracy depends on the provided technology-node. As a rule of thumb, a RC-integrator is a better choice for meeting a linearity of better than 50 dB to 60 dB [35].

into an integrator gain error G_E and a parasitic pole s_{LP}:

$$G_E = \frac{2\pi \cdot GBW}{\frac{1}{RC} + 2\pi \cdot GBW},$$

(2.39)

$$s_{LP} = \frac{C}{C + C_P} \cdot \left(\frac{1}{RC} + 2\pi \cdot GBW\right).$$

(2.40)

Taking these two equations for the real integrator leads to the following high-level description:

$$H_{real} = \frac{1}{s \cdot RC} \cdot \frac{G_E}{1 + \frac{s}{s_{LP}}}.$$

(2.41)

Another important nonideality is the limited output swing of the OpAmp. In a practical implementation of an OpAmp, the gain is not the same for all values of the output voltage. Instead, it decreases as the output voltage increases. This is due to the reduction of the output resistance as the drain-source voltage of the output transistors decreases. The nonlinear gain of the OpAmp can be modeled by a truncated Taylor expansion. The coefficients for the polynomial function can be derived from high-level simulations describing the tolerable nonlinear input-output transfer characteristic.

2.3.6 Design Methodology for CT-Modulators

Continuous-time delta-sigma modulators are often employed as A/D-converters. These modulators are an attractive approach to implement high-speed converters in VLSI systems due to the benefits of a feedback configuration they have low sensitivity to circuit imperfections compared to other solutions such as pipeline- or flash-converters. This section gives a short overview about the analysis, modeling and design of high-speed continuous-time delta-sigma modulators. The resolution and the stability of these modulators are limited by two main factors, excess-loop delay and sampling uncertainty. Both factors, among others, have been carefully analyzed and modeled in [24].

Schreier presents a state-space method of designing ideal continuous-time delta-sigma modulators as described in [56]. Another approach is to find a discrete-time modulator that is equivalent to the desired continuous-time one. Consequently, the differential equations that describe the Continuous Time (CT) modulator are replaced by the difference equations of the Discrete Time (DT) modulator. In [24] this is done by the *impulse-invariant transform*. This approach simplifies enormously the time-domain simulation of the modulator and allows reuse of standard design software for DT systems.

A CT- and a DT-modulator are depicted in Fig. 2.18, on the left hand side and on the right hand side, respectively. The open-loop representation are shown on

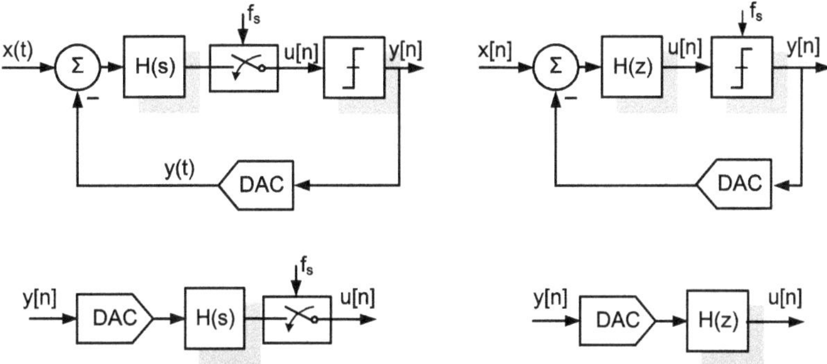

Fig. 2.18 Open-loop representation of CT- (*left*) and DT-modulators (*right*)

the lower side of Fig. 2.18, whereas both input signals $x(t)$ and $x[n]$ are nulled ($x(t) \equiv 0$, $x[n] \equiv 0$). According to [24], a DT delta-sigma modulator is equivalent to a CT one, if the output sequences $y[n]$ of both modulators are identical, assuming that they operate at the same sampling frequency f_s and that the same input signal is applied to both of them. The modulator's output sequence is defined by the quantizer input sequence $u[n]$. Hence, the output sequences $y[n]$ of both systems will be identical if their quantizer-input $u[n]$ sequences are identical as well. Note that for both CT and DT modulators, the input and output of the open-loop configurations are both DT quantities. Assuming the same sequence $y[n]$ is applied to the input of both open-loop systems, the condition for equivalence of both modulators is that the quantizer input sequences $u[n]$ are identical at the sampling instants. This condition is satisfied if the loop transfer functions of the two modulators comply with [24]:

$$Z^{-1}\{H(z)\} = L^{-1}\{H_{DAC}(s) \cdot H(s)\}\big|_{t=nT_s}, \tag{2.42}$$

where $H_{DAC}(s)$ is the transfer function of the D/A-converter. In the time domain equation (2.42) can be expressed as:

$$h[n] = \{h_{DAC}(t) * h(t)\}\big|_{t=nT_s}. \tag{2.43}$$

This transformation between the CT and DT is called the *impulse-invariant transform* because it is required that the open-loop impulse response to be the same at sampling instants $t = nT_s$. A discrete-time loop-filter $H(z)$ can be designed for a delta-sigma modulator by employing discrete-time domain synthesis methodologies and simulations to evaluate the overall performance. The impulse-invariant transform (2.42) can be used to find the continuous-time loop-filter $H(s)$ by taking into account the specific pulse-shape $h_{DAC}(t)$ of the D/A converter and its related $H_{DAC}(s)$.

Considering Fig. 2.18 once again the CT modulator on the left side. The quantizer comprises one or several latched comparators [71] whose output signal $y[n]$ drives the DAC to get the CT representation of signal $y(t)$. Ideally, $y(t)$ responds immediately to the sampler clock edge, but in practice, all involved components in the latch and the DAC cannot switch instantaneously. Thus, there exists a delay between the quantizer clock and the DAC output signal, which is called Excess Loop Delay (ELD). This effect is not present in DT modulators, since it acts as a fully sampled system. The ELD is dependent on many parameters, such as the amplitude $u[n]$ of the input signal to the quantizer, the power supply voltage, the process parameters and the operative temperature. If the ELD extends a certain limit, the closed-loop modulator can easily become unstable [24]. One possible solution is to go for a fixed ELD given to a certain portion of the clock period which needs to be fulfilled by the circuit design. The fixed ELD can be accomplished by using delayed latches clocked by a delay line. During the choice and design of the loop-filter the tolerated ELD has to be taken into account accordingly for achieving sufficient stability margin.

The design methodology proposed in [24] consists of several steps, as follows:

- Design of a $NTF(z)$ that meets the specification (SNR, SNDR, SFDR, DR)
- Calculation of $H(z)$ by means of the linear model
- D/A-pulse type selection and $H(s)$ calculation by means of impulse-invariant transformation (2.42)
- Architecture selection and computation of loopfilter coefficients
- Transient simulation of the architecture on SPICE-level, including linear and non-linear effects
- $H(z)$ extraction from transient simulation by means of impulse-invariant transformation, circuit parameters computation and tuning.

2.3.7 Summary of Delta-Sigma Converter

As a summary, a general formula for the maximum SQNR for a delta-sigma modulator is briefly discussed [56]. This formula gives an overview of the various architectural parameters and their effect on Signal-to-Quantization-Noise Ratio (SQNR):

$$SQNR = \frac{3}{2} \cdot \frac{2L+1}{\pi^{2L}} \cdot \left(2^N - 1\right)^2 \cdot OSR^{2L+1} \tag{2.44}$$

where L stands for the order of the loop filter, N is the number of bits in the quantizer and OSR is the oversampling ratio. Normally, a delta sigma modulator is dimensioned such that its quantization noise within the band of interest is lower than its circuit noise. The circuit noise provides dithering and reduces the appearances of tones and therefore the necessary condition is fulfilled for the linear noise model of a quantizer. Furthermore, this approach yields lower total power consumption. While a reduction of the noise power of an analog circuit requires a strong increase in power consumption, the quantization noise can be lowered by carefully architectural measures in a more power-effective way. In industrial solutions, the value of the oversampling factor is often dictated by the system. The sample-rate might be standardized or a specific clock-frequency is available and should be reused to keep the overall costs low. For a fixed clock-frequency, a possibly high filter-order should be implemented, because of two reasons. First, the high-order filter stages provide additional shaping of the quantization noise. Second, the power drain of these stages can be kept low, since the filter is put inside of an overall feedback loop and the noise and distortion of these stages is suppressed by the gain of the preceding stages. On the other hand, stability issues put an upper limit on the filter order. Multi-bit quantization provides increased resolution without increasing the filter-order or sample-rate. This translates in a higher maximum stable input range, lower quantization noise and a well defined quantizer gain as compared to a single-bit design. However, single-bit quantizers have the advantage of inherent linearity while conventional multi-bit designs need DEM of the unit-elements in the feedback D/A-converter [56] and [54]. The implementation of the DEM-algorithm results in an increased circuit complexity and increased delay in the feedback path. In a summary, multi-bit modulators show improved stability behavior while single-bit designs show well traded power over performance ratio for most applications, unless the band of interest goes to very high frequencies compared to the speed of the provided technology node.

2.4 Asynchronous Delta-Sigma Modulation

Modern CMOS technologies employ the high speed performances of nanoscale transistors. Maintaining adequate reliability over the required lifetime reduces the possible maximum supply voltage. Commonly, the required lifetime is in the range of 15 years for a product in the field of telecommunications. The possible supply voltage results in approximately 1 Volts for a device consisting of a gate-oxide thickness of about 1.2 nm. These low-voltage CMOS technologies pose new issues in the design of multi-bit delta-sigma converters. One of the main challenges is the implementation of the quantizer with an array of comparators resulting in a flash-converter. The performance of these converters is bounded by the tolerable limit for the power consumption of the flash-converter embedded in multi-bit architectures. This flash-converter is quite problematic to realize in low-voltage technologies, where the dynamic-range of the comparators is limited and comparator offset degrades the linearity of the quantizer. On the other hand, the decrease in the minimum feature size has a beneficial effect on the obtainable time resolution in the circuitry, owing to the increase in the intrinsic speed of the transistors.

Some alternatives have been explored to avoid use of direct amplitude quantization in A/D-converters, which are based on *time encoding* rather than *amplitude encoding* [9]. An exchange of the amplitude-axis for the time-axis offer a possibility of overcoming resolution problems in A/D-conversion in low-voltage CMOS circuits. This exchange can be done by some form of *duty-cycle modulation* or *pulse-width modulation*. Conceptually, this would mean an amplitude to square-wave conversion by some square-wave modulation scheme, followed by a discretization of the time-axis. For its implementation a circuit configuration is used consisting of an *asynchronous* delta-sigma modulator. This approach converts an analog time-continuous

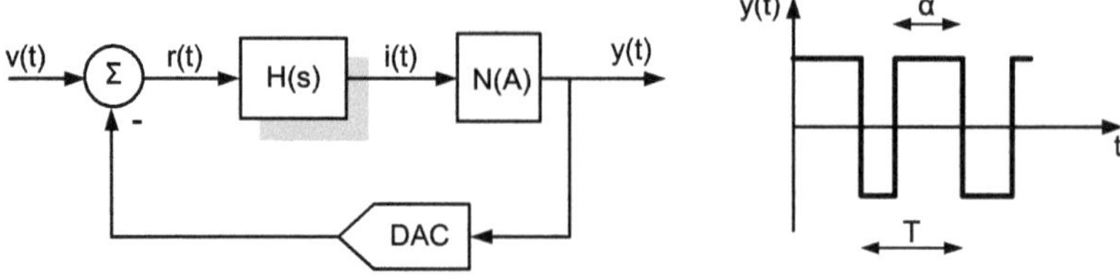

Fig. 2.19 Block diagram of an asynchronous delta-sigma modulator

input signal into a time-continuous discrete amplitude output signal. Asynchronous delta-sigma modulators consists of the same configuration as their synchronized counterparts, namely the previously described delta-sigma converters. Respectively, they are a closed-loop, nonlinear system comprising a linear filter-stage $H(s)$ and a non-linear element with a dedicated transfer function $N(A)$ dependent on the input amplitude [57]. Figure 2.19 depicts a block diagram of an asynchronous modulator and the waveform at the output with period T and duty-cycle $\frac{\alpha}{T}$. In the most general case $N(A)$ can be any type of quantizer, but preferably would be a simple comparator.

Note, that unlike in conventional delta-sigma modulators, no sampling operation is performed and consequently no quantization noise is introduced into the system. However, the transformation of amplitude into time information is done via an inherent limit-cycle of the delta-sigma operation. Obviously, without any high-frequency signal inside the loop, the modulator output signal would be hold on a constant value and no signal conversion could occur. In conventional delta-sigma modulators any occurring limit-cycles or idle-tones are treated as an undesired effect, as discussed in [56]. In an asynchronous modulator the limit-cycle is a required effect for generating a high frequency signal to enable duty-cycle modulation or pulse-width modulation.

This book concentrates on asynchronous delta-sigma modulation employing a single, stable and unforced limit-cycle using a single-bit quantizer to achieve a low cost solution in terms of power consumption and chip area. The term oscillation frequency (ω_c) will be used for the oscillation frequency of the limit-cycle. The system can be characterized as proposed in [9] by means of the duty cycle $\frac{\alpha}{T}$, whereby α is a pulse-width and T is a time-variant period. The instantaneous frequency ω of the output signal can be expressed as:

$$\frac{\alpha}{T} = \frac{v+1}{2}, \qquad \frac{\omega}{\omega_c} = 1 - v^2, \tag{2.45}$$

where v stands for the amplitude of $v(t)$, normalized to one, $|v| \leq 1$ and $\omega = \frac{2\pi}{T}$. The information is coded into the zero crossings of the output two-level square wave. If the limit cycle oscillation has a much higher frequency than the modulating input signal, then the spectral component of the baseband signal at the output is fully preserved.

2.4.1 *Spectral Analysis of Duty-Cycle Modulation*

The limit-cycle oscillation in a closed-loop non-linear system will be evaluated in terms of the existence, stability and frequency. There is a dependency on the properties of both kind of elements, the linear and non-linear. Such a system has been analyzed in [9] among other publications in the field of pulse-width modulation. This book focuses on a single-bit quantizer with hysteresis claiming that this approach gives a cost-effective implementation. The nonlinear element $N(A)$ in Fig. 2.19 will be assumed to be a simple comparator with hysteresis for further investigations.[4] The resulting system can be analyzed by deriving an exact analytical solution by employing a Fourier-series expansion of the square-wave output signal. This leads to an explicit determination of the limit-cycle oscillation frequency and gives an expression for the spectral content in the loop, when the asynchronous delta-sigma modulator is modulated by the input signal. Under the initial assumption of a zero input signal with 50 % duty-cycle, the resulting square wave signal $y(t)$ is expanded in a Fourier-series:

$$y(t) = \frac{4}{\pi} \sum_{k=1,3,5,\dots}^{\infty} \frac{\sin k\omega_c t}{k} \tag{2.46}$$

whereby the output level of the quantizer is normalized to one and ω_c is the frequency of the square-wave. The input signal $i(t)$ of the quantizer can expressed as a convolution of the filter impulse response and the residue signal $r(t) = v(t) - y(t)$ (see Fig. 2.19). Furthermore, it is taken into account that the instants when $i(t)$ crosses the hysteresis values h and $-h$, correspond to the zero crossings of the square-wave. The solution of the resulting system of equations for a quantizer with hysteresis h leads to following set of criteria:

$$\sum_{k=1,3,5,\dots}^{\infty} \Re\left\{H(jk\omega_c)\right\} < 0. \tag{2.47}$$

This gives the condition under which the above assumptions are going to be fulfilled and with

$$\sum_{k=1,3,5,\dots}^{\infty} \frac{1}{k} \cdot \Im\left\{H(jk\omega_c)\right\} = -\frac{\pi h}{4} \tag{2.48}$$

the exact frequency of the limit cycle for a particular filter implementation can be determined.

The spectral properties of an asynchronous delta-sigma modulator is done in [9] via an evaluation of the transformation: amplitude into time. The spectral properties of the pulse-width modulated output signal is done as described above, but applying

---

[4]A hysteresis of $\pm h$ is assumed for the considered comparator.

Fig. 2.20 A single-bit asynchronous delta-sigma modulator

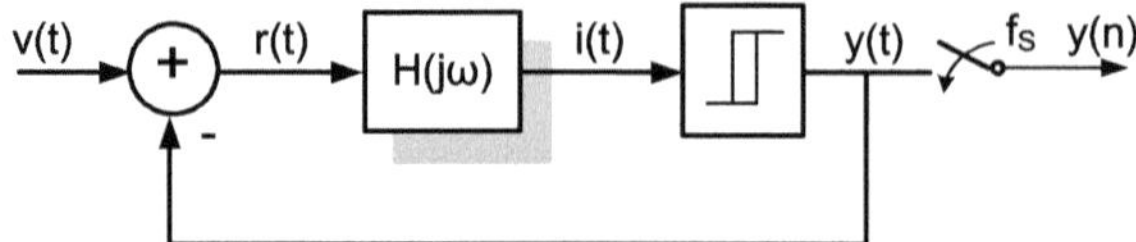

a harmonic input signal $v(t) = v_m \cos \mu t$. The spectral components in the first harmonic band around the center-frequency due to modulation for a sinusoidal input signal can be expressed as:

$$y_1(t) = \frac{2}{\pi} \sum_{-\infty}^{+\infty} \sum_{-\infty}^{+\infty} J_m\left(v_m \frac{\pi}{2}\right) \cdot J_{2n}\left(\frac{\omega_c v_m^2}{4\mu}\right) \cdot \cos m\frac{\pi}{2} \cdot \cos\left(\omega_0 t + (2n + m)\mu t\right)$$

(2.49)

with

$$\omega_0 = \left(1 - \frac{v_m^2}{2}\right)\omega_c$$

whereby, in general, $J_p(x)$ denotes a Bessel function of the first kind and order p. High frequency components are introduced into the spectrum of the modulated signal. The first harmonic band, given by $y_1(t)$, contains spectral components around a center-frequency ω_0 whose amplitudes are Bessel functions in the case of harmonic modulation $v(t) = v_m \cos \mu t$. For a good design, the center frequency ω_0 must be high enough to separate the significant components from the baseband. Note, the center frequency shifts as a function of the amplitude of the modulating input signal $v(t)$ relative to the carrier-frequency ω_c. It can be concluded that the harmonic components are increasing as a function of the input amplitude. Thus ω_c has to be chosen high enough to avoid any distortion components in the band of interest due to formula (2.49). Thereby, any distortion remains below the noise floor even for high input values for amplitude and frequency. The baseband equals the modulating input signal $v(t)$ without distortion.

2.4.2 An Asynchronous Delta-Sigma Modulator

Figure 2.20 shows the scheme of an asynchronous delta-sigma modulator using a simple comparator as presented in [9]. It is shown in [9] that this configuration is an proper implementation of the duty-cycle modulation and can be used as an implementation for an asynchronous delta-sigma modulator. Unlike in the case of synchronous or conventional delta-sigma modulation, the sampler is configured outside the loop. Thereby, no quantization noise occurs inside the loop and consequently no high frequency components are inserted into the loop. This fact makes the implementation more robust and easier to design than in a high frequency delta-sigma converter.

Note, the signal at the output of the comparator $y(t)$ is still continuous in time by means of pulse-width modulation. Digital signal processing requires a discretization

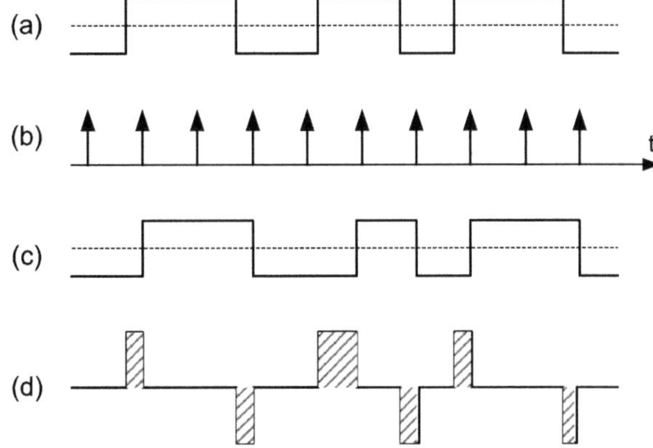

Fig. 2.21 Sampling of a PWM-signal

in time which is done by a subsequent sampler. The sampling process of a pulse-width modulated signal introduces a certain error, as illustrated in Fig. 2.21, due to the fact that the leading and trailing edges are not exactly aligned with a train of equidistant sampling pulses.

In Fig. 2.21(a) signal $y(t)$ depicts the modulated asynchronous square-wave. Figure 2.21(b) shows a sequence of equidistant sample pulses for the sampling process, Fig. 2.21(c) reveals the waveform after sample and hold and Fig. 2.21(d) gives the derived error signal due to equidistant sampling. The lower the sampling-rate, the higher the mismatch between the original signal and the waveform after sampling, i.e. the higher the quantization noise as represented by the difference signal. In [9] the total quantization noise power P_N within a certain bandwidth BW is derived and amounts to

$$P_N = \frac{8}{3} \cdot f_0 \cdot T_s^2 \cdot BW, \qquad f_0 = \frac{\omega_0}{2\pi} \tag{2.50}$$

whereas T_s stands for the sampling period and f_0 is the center-frequency. This result is based on the assumption that the area of the error signal is a random quantity and thus there is no correlation between the sampling frequency and the modulating frequency. This straightforward discretization of the time-axis requires a very high sampling-frequency to achieve the desired dynamic-range. The high speed requirements on the integrated-circuit seems to make this approach less attractive. Nevertheless, the offered high-speed features in nanoscale technologies put a lot of attractiveness on this proposed modulation scheme. Furthermore, the demodulation of a pulse-width modulated signal is carried out by a time-to-digital decoder to convert the PWM signal into a uniformly sampled signal. This decoder is based on the fact that the input signal spectrum is still present in the output spectrum of the PWM signal. The digital decoder is realized by a poly-phase sampler [9] or by a time-to-digital converter as proposed in [19] or by applying fast recovery algorithms for a time encoding machine (TEM) as discussed in [1]. A time-to-digital converter measures the pulse-widths of the modulated signal and a digital reconstruction circuit translates the pulse-widths back to the original amplitudes. This is accomplished in three steps:

- Firstly, measurement of the edge position of the PWM-signal relative to the rising edge of a clock reference. The measured results are updated synchronously with the reference clock. To be able to detect all edges, the reference clock must be bigger than the maximum pulse rate, i.e. the center frequency.
- Secondly, the pulse-widths and their positions are calculated digitally based on the measurements.
- Finally, the original signal can be reconstructed by employing the derived results such as high-time and low-time of pulses and position of edges in time.

The next interesting point is the investigation on distortion in an asynchronous delta-sigma modulator. The most significant distortion term reveals to be the third harmonic component. The definition of a distortion coefficient Δ_3 as the ratio of the fundamental component and the third harmonic can be expressed as:

$$\Delta_3 = \frac{\pi^2}{6} \frac{\Re\{H(jk\omega_0)\}}{H(\mu)} v_m^2. \tag{2.51}$$

A general nth order loopfilter $H(\omega)$ with one pole p_1 of nth order and one zero z_1 of $(n-1)$th order can be expressed as:

$$H(\omega) = \frac{(j\omega + z_1)^{n-1}}{(j\omega + p_1)^n} \frac{p_1^n}{z_1^{n-1}}. \tag{2.52}$$

Taking this loopfilter for calculating the general third harmonic component leads to:

$$\Delta_3 = \frac{\pi^2}{6} \frac{\mu^n}{\omega_0^n} v_m^2 = \frac{\pi^2}{6} \frac{\mu^n}{\omega_c^n} \frac{v_m^2}{(1 - \frac{v_m^2}{2})^n} \tag{2.53}$$

and thus the size of the third harmonic distortion depends on the order of the loopfilter, the ratio $\frac{\omega_c}{\mu}$ and the amplitude v_m of the input signal.[5]

In Fig. 2.22 an output spectrum is shown of a PWM-modulator as depicted in Fig. 2.20. The center-frequency was set to 10 MHz and the applied harmonic signal has a sinusoidal waveform with an amplitude of -6 dBFS and a frequency of 1 MHz. Full-scale (FS) is referred to the comparator output value. The distortion components due to the modulation is centered around 8.75 MHz and the distortion components due to the Bessel functions are visible down to 3 MHz. The tolerable input amplitude is limited due to the fact that an increase of the input amplitude causes a raise of distortion components. In the shown example, the inband distortion is negligible.

---

[5] A harmonic signal $v(t)$ is considered as input signal $v(t) = v_m \cos(\mu t)$.

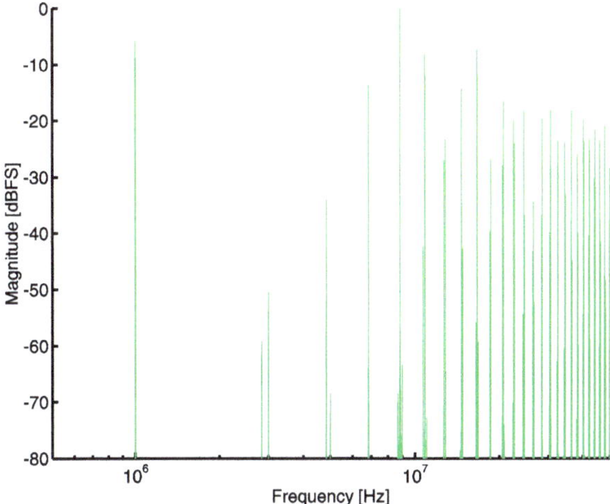

Fig. 2.22   Spectrum of PWM-modulation

2.4.3 Summary of Asynchronous Delta-Sigma Modulation

Asynchronous and synchronous delta-sigma modulation employs the principle of
pulse-modulation characterized by the use of an analog reference input to the modulator. In general, pulse-modulation systems represent a message bearing a signal
by a train of pulses. Among different basic pulse-modulation techniques the *Pulse-
Density-Modulation* and the *Pulse-Width-Modulation* are used for synchronous and
asynchronous delta-sigma modulation, respectively. The Pulse-Density-Modulation
is based on an unity pulse-width, height and a constant time of occurrence for the
pulses within the switching period. The modulated parameter is the *presence* of the
pulse. For each sample interval it is determined if the pulse should be present or not,
hence the designation *density* modulation. In contrast, the Pulse-Width-Modulation
codes the information into the pulse time position within each switching interval.
The *width* of each pulse represents an amplitude value at a particular instant. The
coded information in the time domain needs to be converted into the digital domain
by means of sampling of the pulse-width signal by uniform time intervals. The resulting signal is discrete in time and amplitude and can be applied to further digital
signal-processing. The spectral analysis reveals the individual components present
in the output spectrum. The modulating input signal is left unchanged by the pulse-
width modulation. There are no direct forward harmonics of the input signal, i.e.
the pulse-width modulation process can be considered *ideal* in terms of harmonic
distortion. However, the output spectrum shows discrete tones of the harmonics
of the carrier-frequency and intermodulation products of the input frequency and
the carrier-frequency. The whole systems comprises a pulse-width modulator followed by a digital demodulator offering convenient implementation characteristics
in low-voltage nanoscale technologies. The quantization in amplitude is exchanged

by quantization in time to achieve high resolution. This approach is in favor with the very high speed features of nanoscale devices.

2.5 Impact on Jitter

One key factor for high performance A/D-converters is the availability of very low jitter sampling clocks. The jitter of a clock is the displacement of a real clock-edge referred to a desired instant in time leading to a random deviation of the clock frequency. This can be seen as random variation in the period or deviation of the zero crossing points from their ideal position along the time-axis. For recent wideband applications such as WLAN, clocks with jitter in the order of few picoseconds are required. The term *jitter* is used in the time domain, the same phenomenon can be characterized in the frequency domain and is expressed by *phase noise*. For a periodic sinusoidal signal $x(t) = A\cos(\omega_c T + \phi_n(t))$, where $\phi_n(t)$ is a small random excess phase representing variations in the period. The function $\phi_n(t)$ is called *phase noise*. For an ideal sinusoidal signal operating at ω_c, the spectrum assumes the shape of an impulse, whereas for a real signal, the spectrum exhibits "skirts" around the carrier frequency ω_c, as described in [5].

2.5.1 Jitter Definition

An important aspect for the sampling process is how to specify requirements on the jitter of the clock provided for A/D-conversion. In the available literature, there are a number of different jitter definitions such as *cycle, cycle-to-cycle, period, accumulated, absolute, long-term*. Furthermore, for each of those definitions different values can be provided such as the rms-value, 3-sigma value, peak-value or peak-to-peak value. Following definitions are given which are relevant for the sampling process, as shown in [38].

- *Absolute Jitter* (σ_{abs}): The absolute jitter is measured against an ideal clock reference. The trigger is set on a rising edge of the ideal clock and the time differences are measured between the real and ideal clock edges.
- *Period Jitter* (σ_{per}): The period of several clock cycles are measured to show the time variation of the clock cycle. This gives the minimum and maximum clock cycle which is important for synthesis constraints in digital timing.
- *Accumulated Jitter* ($\sigma_{acc}(m)$): This is the time displacement of the clock edges relative to a starting (triggering) edge of the same clock. It is a function of the number of clock cycles which elapse between the triggering edge and the measured edge. Thus, it is a function of discrete time events and can be expressed as $t_j(kT + mT) - t_j(kT)$, whereas t_j denotes the actual jitter on the sampling clock. The edge at time instant kT is the triggering event and the jitter is measured on an edge arising after m cycles.

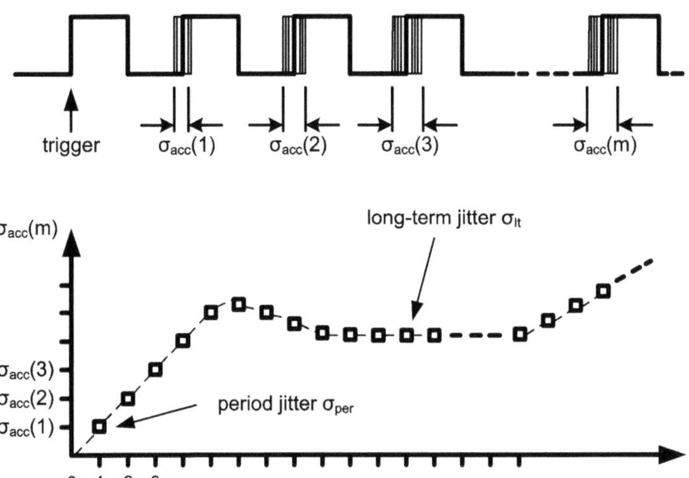

Fig. 2.23 Accumulated jitter of a PLL

- *Long-term Jitter* (σ_{lt}): In a Phase-Locked Loop (PLL) the accumulated jitter function settles around a given value for reasonably large values of the event index m. This number is called the long-term jitter is indicated as σ_{lt}.

The definition of long-term jitter will be further clarified. If $\sigma_{acc}(m)$ of a free running oscillator is measured, the result will be a steadily increasing function, since in autonomous oscillators the intrinsic jitter gets accumulated without showing any settling behavior. If the same oscillator is used in a PLL for building a frequency synthesizer, the accumulated jitter measured at the output of the PLL will result in a characteristic as shown in Fig. 2.23. The accumulated jitter function starts rising for small number m of clock-cycles before it settles around a given value for reasonably large values of m. The settling behavior depends strongly on the relative noise levels of the PLL's input clock and on the PLL's internal noise sources. For very large values of m, the $\sigma_{acc}(m)$ will begin to ramp again, since the jitter of the PLL's input clock becomes dominant. Nevertheless, this second ramp has no impact on a sampling process, because any variation over a very large number of m can be treated as a PLL wander of the output clock [38]. Obviously, following relationship is valid: $\sigma_{acc}(1) = \sigma_{per}$. Note, the measurement procedure defines the *type* of jitter whereas the chosen postprocessing of the measured jitter defines the *statistical parameter* of jitter. For practical reasons, the rms-value is used for expressing the accumulated jitter.

2.5.2 SNR vs. Jitter of Sampled Signals

In the available literature, the jitter of the sampling clock is considered to be Gaussian-distributed showing a white phase-noise and thus flat spectrum. This as-

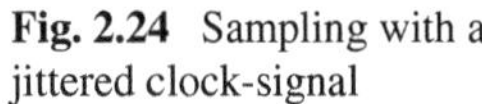

Fig. 2.24 Sampling with a jittered clock-signal

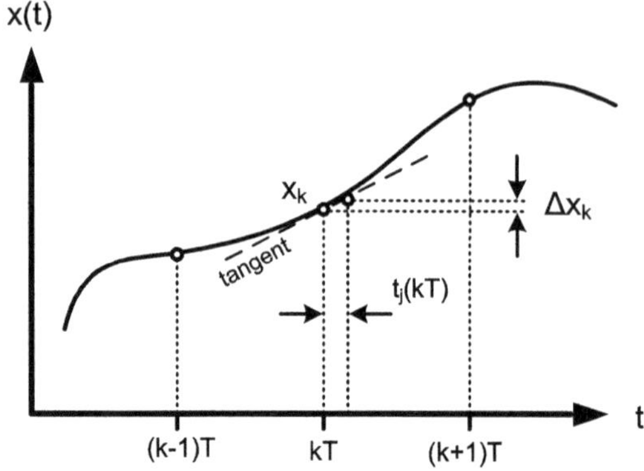

sumption is not applicable in real system-on-chip solutions, since an implemented PLL reveals a lowpass-shaped spectral distribution of the phase-noise. The impact on the SNR due to sampling instant uncertainty will be investigated. The sampling period T and the jitter t_j is considered as time discrete random process with time step T, showing zero mean value and being independent from the analog time continuous input signal $x(t)$. The presence of jitter t_j on the sampling clock, the instants of sampling are displaced from the ideal point. That means, the kth sample will not be sampled exactly at time instant kT, but at $kT + t_j(kT)$. Deriving the expression for the SNR, the *autocorrelation function* of the sampled signal will be calculated using the linear approximation. Employing the *first-order Taylor expansion* for the linear approximation, as illustrated in Fig. 2.24, replaces the ideal sampled value x_k by the real sampled value $x_k + \Delta X_k$ lying on the tangent to $x(kT)$ at the displacement in time at $kT + t_j(kT)$.

As shown in [38], the SNR is calculated by the ratio of the signal power to the error power caused by the jittered clock-signal, knowing that the signal power is the value of the autocorrelation of the input signal $x(t)$ at zero denoted by $r_x(0)$:

$$SNR = 10 \cdot \log_{10}\left(\frac{r_x(0)}{-r_x''(0) \cdot r_{tj}(0)}\right) \text{ [dB]} \tag{2.54}$$

whereby $r_x''(0)$ denotes the second-time derivative of the autocorrelation function of $x(t)$ at zero and $r_{tj}(0)$ is the autocorrelation function of t_j at zero. The achievable SNR due to a jitter on the clock-signal for the A/D-conversion can be calculated for any kind of input signal $x(t)$ and any type of jitter t_j. Expression (2.54) quantifies the intuitive fact that the sampling of slowly varying input signals is less affected by the jitter of the sampling clock. Note, the SNR is independent from the autocorrelation function of the jitter, as long as $r_{tj}(0)$ remains the same. Thus, in the frequency domain the SNR is independent from the spectral distribution of jitter, as long as the total power of the spectrum $r_{tj}(0)$ remains constant.

The relevant jitter specification for A/D-applications can be derived from the autocorrelation function in zero ($r_{tj}(0)$). The measurement of $r_{tj}(0)$ can be done in the frequency domain as well as in the time domain.

- Measurement of $r_{tj}(0)$ in the *frequency domain*: It is common practice to measure a phase noise spectrum $\phi(t)$. The relationship between jitter and phase noise is

$$t_j(t) = \frac{\phi(t)}{2\pi} \cdot T \tag{2.55}$$

where T is the period of the clock. Thus, the power spectra of $\phi(t)$ and $t_j(t)$ are related by following equation:

$$S_{tj}(f) = \left(\frac{T}{2\pi}\right)^2 \cdot S_\phi(f). \tag{2.56}$$

By definition of power spectrum, the value $r_{tj}(0)$ is equal to the integral of $S_{tj}(f)$ from $-\infty$ to $+\infty$. The practical limitations on spectrum analyzers put less attention on measurement of phase-noise for obtaining a measurement result of clock jitter

- Measurement of $r_{tj}(0)$ in the *time domain*: The accumulated jitter can be measured using proper instrumentation with sufficient accuracy and is thus more suitable than the phase noise for the evaluation of $r_{tj}(0)$. The variance $\sigma^2_{acc}(m)$ of accumulated jitter can be expressed in terms of the autocorrelation of the jitter process as follows [38]:

$$\sigma^2_{acc}(m) = 2 \cdot \left[r_{tj}(0) - r_{tj}(mT)\right]. \tag{2.57}$$

Thus, the measurement of accumulated jitter reveals the autocorrelation function of the jitter random process.

A practical PLL contributes a lowpass-shaped phase noise spectrum and thus for large value of t the $r_{tj}(t)$ equals zero. By taking the limit towards infinity on both sides of (2.57)

$$r_{tj}(0) = \frac{\sigma^2_{acc}(\infty)}{2} = \frac{\sigma^2_{lt}}{2}. \tag{2.58}$$

Applying formula (2.54) for an sinusoidal input yields following relationship. Assuming $x(t) = A \cdot \sin(\omega t)$, the autocorrelation function is

$$r_x(mT) = \frac{A^2}{2}\cos(\omega m t) \tag{2.59}$$

and the second derivative is

$$r_x''(mT) = \frac{-\omega^2 A^2}{2}\cos(\omega m t). \tag{2.60}$$

By substituting in formula (2.54) the well known formula results for aperture jitter SNR in sampling of sinusoidal signals:

$$SNR_j = 20 \cdot \log_{10}\left(\frac{1}{\omega \sigma_{tj}}\right) \; [\text{dB}] \tag{2.61}$$

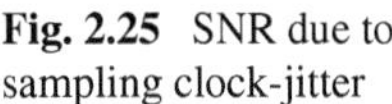

Fig. 2.25 SNR due to sampling clock-jitter

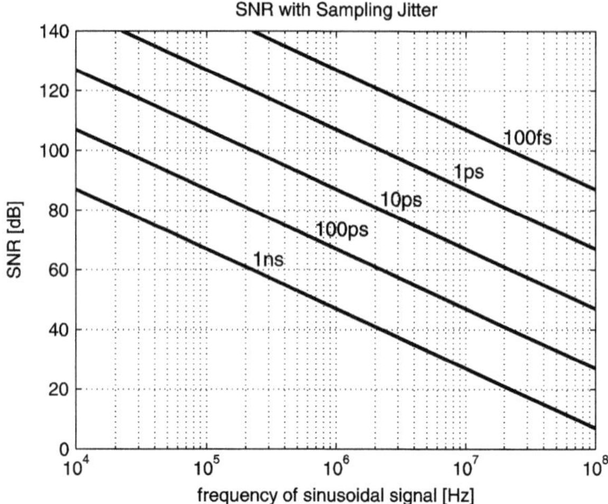

whereas σ_{tj} is the rms-value of the stochastic process causing clock jitter. Taking into account the long-term rms jitter of a conventional PLL, formula (2.61) in conjunction with formula (2.58) then becomes:

$$SNR_j = 20 \cdot \log_{10}\left(\frac{\sqrt{2}}{\omega \sigma_{lt}}\right) \text{ [dB]}. \tag{2.62}$$

Figure 2.25 shows the obtainable SNR as a function of the signal frequency according to (2.62) with σ_{lt} as a parameter.

2.6 Conclusion

This chapter dealt with the basics of A/D-conversion focusing on delta-sigma modulation. The conversion of analog signals is accomplished by sampling in time and quantization in amplitude. Fulfilling the *Nyquist theorem* (2.1) allows a reconstruction of the original analog signal by simple lowpass filtering, as shown in Fig. 2.1. The *white-noise model* (2.4) for describing the quantization error was introduced to simplify the analysis of A/D-converters. The most important specifications were briefly discussed, whereas the dynamic performance measures are more relevant for delta-sigma modulators. The dynamic performance can be expressed by Signal-to-Noise Ratio (SNR) (see formula (2.6)), Total Harmonic Distortion (THD), Signal-to-Noise and Distortion Ratio (SNDR), Dynamic Range (DR), Spurious-Free Dynamic Range (SFDR), Idle Channel Noise (ICN) and the overload level, as depicted in Figs. 2.4 and 2.5.

Delta-Sigma modulation is a technique which combines filtering and oversampling to perform analog-to-digital conversion. Figure 2.8 shows all relevant components comprising a whole A/D-conversion path. The quantization noise from a low resolution quantizer is shaped away from the signal band prior to being removed by proper filtering. The modulator is described by the Signal Transfer Function (STF) and Noise Transfer Function (NTF) in the z-domain or s-domain. The In-Band Quantization Noise (IBQN) can be approximated by formula (2.17) as illustrated in Fig. 2.11. The architecture can be chosen for a single-loop or cascaded-loop approach trading-off stability problems against noise leakage. The loopfilter can be composed by employing discrete-time or continuous-time filter such as switched-capacitor circuits or active RC-filters and $g_m C$-filters, respectively. Both approaches reveal nonidealities deteriorating the overall performance such as finite OTA-gain, finite gain-bandwidth-product and limited slew-rate of the OpAmp as well as parasitic capacitances and resistances. The impact of these nonidealities needs to be well understood and a proper modeling is required to derive a well balanced block-specification for achieving the targeted converter performance. The $\Delta\Sigma$-modulator's Signal-to-Quantization-Noise Ratio (SQNR) is defined by the order of the loopfilter, the number of bits in the quantizer and the Over-Sampling Ratio (OSR), as described in formula (2.44).

An exchange of the amplitude-axis for the time-axis offer a possibility of overcoming resolution problems in A/D-conversion in low-voltage CMOS circuits. This exchange can be done by some form of *duty-cycle modulation* or *pulse-width modulation*. The spectral analysis of duty-cycle modulation reveals an undistorted input signal in the baseband as well as high frequency components dominated by the first harmonic band around a center frequency whose amplitudes are Bessel functions in the case of harmonic modulation, as shown in formula (2.49) and Fig. 2.22. Figure 2.20 shows the scheme of an asynchronous delta-sigma modulator using a simple comparator. This configuration is an proper implementation of the duty-cycle modulation and can be used as a realization for an asynchronous $\Delta\Sigma$ modulator.

One key factor for high performance A/D-converters is the availability of very low jitter sampling clocks. The obtainable SNR can be expressed by formula (2.62) as a function of the signal frequency according to the long-term jitter (σ_{lt}) as a parameter. The accumulated jitter can be measured using proper instrumentation with sufficient accuracy. The long-term jitter can be derived from the accumulated jitter curve which is relevant for the sampling process revealing the achievable SNR as depicted in Fig. 2.25.

Chapter 3
A Delta-Sigma Converter with Dynamic-Biasing Technique

A high-resolution multi-bit Delta-Sigma ADC implemented in a 0.18 μm CMOS technology is introduced [47]. The circuit is targeted for an ADSL Central-Office application. An area- and power-efficient realization of a single-loop modulator consisting of a 2nd-order loopfilter and a 3-bit quantizer with an oversampling-ratio of 96 is presented. The delta-sigma modulator features an 85 dB dynamic-range (DR) over a 300 kHz signal bandwidth. The measured power consumption of the ADC core is 15 mW only. An innovative biasing circuitry is introduced for the switched-capacitor integrators. The FOM taking the DR as reference (2.7) results in $1.8 \frac{\text{pJ}}{\text{conv}}$.

The growing market for Asymmetric Digital Subscriber Line (ADSL) applications is the driving force for increasing the integration level of such a transceiver Application-Specific Integrated Circuit (ASIC). Digital design demands deep submicron technologies in order to reduce power consumption and to enlarge the integrated functionality. For analog modules, migration to supply voltages below 3.3 V is more critical. Multi-chip packaging is not always a preferable approach since the total manufacturing costs are quite high. Taking these points into account it becomes necessary to build an Analog Front End (AFE) in a 0.18 μm technology, allowing the integration of the AFE together with digital blocks of the modem on one single silicon die.

As illustrated in Fig. 1.3, in an ADSL modem (see Fig. 1.2) the downstream data are transferred at higher frequencies in a wider band as compared to the upstream data offering different data-rates, as shown in Table 1.1. Flexible band allocations are standardized, such as frequency division (FDD), frequency overlapping, single upstream and double upstream, as already described in the first chapter. The developed converter targets a CO line-card application. For such transceivers, the ADC typically requires 14 bit resolution over an analog signal bandwidth of 300 kHz. The harmonic distortion must not exceed 80 dBc to not compromise the maximum data-rate.

R. Gaggl, *Delta-Sigma A/D-Converters*, Springer Series in Advanced Microelectronics 39,   55
DOI 10.1007/978-3-642-34543-2_3, © Springer-Verlag Berlin Heidelberg 2013

3.1 Architectural Considerations

In line-card applications, the power consumption is one of the most important criteria for the analog front end. A high-power dissipation limits the achievable density factor. This has a strong impact on the architectural considerations. Many published ADSL modulators employ cascaded structures [4, 17, 27, 64, 76]. In general, multi-stage modulators suffer from leakage noise, challenging the design of analog building blocks. Leakage noise is quantization noise mainly caused by the first-stage quantizer, which is not fully canceled in the digital noise-cancellation logic, as described in the second chapter by condition (2.20). Leakage noise caused by analog building blocks depends highly on the accuracy of gain-factors, finite operational-amplifier (OpAmp) DC-gain, finite OpAmp closed-loop pole and on nonzero switch-resistances (see formula (2.24), (2.28) and (2.30)). Since it is hard to design OpAmps with quite high DC-gain in deep sub-micron technologies, cascading topologies were not considered adequate and a single-loop modulator was selected.

The ratio of transconductance g_m over output conductance g_{DS} is proportional to the OpAmp DC-gain. MOS devices in deep sub-micron technologies feature high g_m but also high g_{DS} values. Thus, the OpAmp DC-gain reduces with a decreasing feature size. To make up for this effect, gain-boosting techniques can be applied. However, the design of a gain-boosting stage is not trivial. It has its own gain, own bandwidth and hence own gain-bandwidth-product which has to coincide exactly with the bandwidth of the original cascode amplifier. Any mismatch between both frequencies reveals a pole-zero doublet causing an increased settling time [71]. In switched-capacitor filters, the settling time determines the minimum width of the clock-pulses and therewith the maximum clock-frequency and hence the possible OSR. Therefore, in such a system, pole-zero doublets must be avoided in any case. One possible solution to avoid an auxiliary gain boosting amplifier is going for cascading of amplifier-stages instead of cascoding. The voltage gains are equal but the bandwidths are different, since each high-ohmic node with its capacitance to ground gives a pole. Furthermore, the power dissipation for a two-stage amplifier such as a Miller-OpAmp becomes higher, but the possible output swing is increased which is favorable for low-voltage designs. Deep sub-micron devices also provide low intrinsic parasitic capacitances due to smaller geometrical dimensions. In other words, new technologies offer high-speed devices at the cost of reduced DC-gains.

The Signal-to-Quantization-Noise Ratio (SQNR) of an ideal Lth-order delta-sigma modulator with an oversampling ratio OSR and N-bits in the quantizer can be expressed by formula (2.44). The stability constraints of high-order single-loop topologies were discussed in the second chapter, thus, the order of noise-shaping was chosen to be two. This approach needs a small number of building blocks, hence the necessary silicon area becomes quite small. As a trade off, the OSR was chosen to be 96, resulting in a sampling frequency of 53 MHz. As discussed above, this choice is very much in favor with the high-speed properties of deep sub-micron technologies.

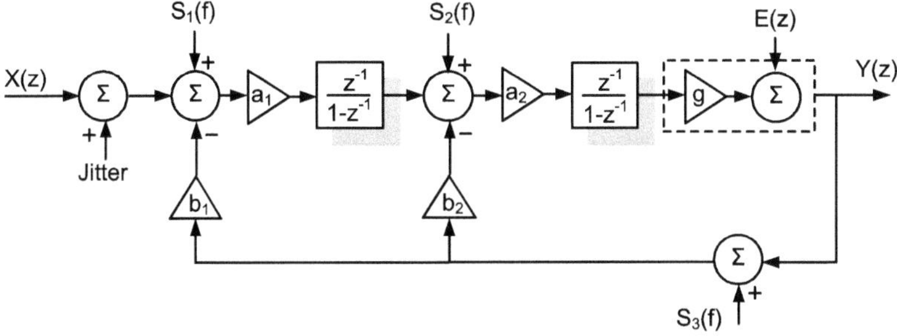

Fig. 3.1 Second-order noise-shaping loop

The implementation of multi-bit quantizers has become a state of the art design. Deep sub-micron processes feature excellent matching characteristics. Thus, careful design and layout can fulfill the linearity requirements on the internal Digital-to-Analog Converter (DAC). For requirements beyond 14 bits Dynamic Element Matching (DEM) techniques can be employed [65]. To bring the quantization noise floor well below the 14 bit level, a 3 bit internal quantizer was chosen. The physical dimensions of quantizer and DAC are significant as compared to the total converter area.

3.1.1 Considerations on Signal-to-Noise Ratio

Figure 3.1 depicts a standard 2nd-order delta-sigma ADC employing multiple feedback. The inner feedback loop (coefficient b_2) keeps the whole modulator stable by providing sufficient phase margin [56]. The sampling frequency of 53 MHz makes a robust switched-capacitor implementation possible in a 0.18 μm CMOS technology. The speed requirements on the OTAs are still meaningful to achieve a low-power design. A switched-capacitor loopfilter can make use of the perfect capacitor matching properties showing a mismatch factor better than 0.1 %. The ADC requirements would not justify to go for a continuous-time loopfilter implementation due to the related nonidealities, as mentioned in chapter two.[1]

The high OSR of 96 allows a placement of the NTF's zeros at DC enabling the usage of two cascaded integrator stages. The proposed second-order modulator employs two delaying integrators which is desirable because it allows the OTAs in each integrator to settle independently of each other, thereby relaxing their speed requirements [54]. The linear transfer function of a discrete-time delaying integrator results in

$$H(z) = \frac{z^{-1}}{1 - z^{-1}}. \tag{3.1}$$

---

[1] Note, for the given bandwidth of 300 kHz a reasonably high GBW can be afforded in the provided technology.

Presuming unity gain g for the multi-bit quantizer, linear analysis leads to following Signal Transfer Function (STF):

$$STF(z) = \frac{a_1 a_2 z^{-2}}{(1-z^{-1})^2 + a_2 b_2 z^{-1}(1-z^{-1}) + a_1 a_2 b_1 z^{-2}}. \tag{3.2}$$

The Noise Transfer Function (NTF) results in

$$NTF(z) = \frac{(1-z^{-1})^2}{(1-z^{-1})^2 + a_2 b_2 z^{-1}(1-z^{-1}) + a_1 a_2 b_1 z^{-2}}. \tag{3.3}$$

In order to achieve $STF(z) = z^{-2}$ and placing both NTF's zeros at DC, that means $NTF(z) = (1-z^{-1})^2$, constrains the choice for coefficients in the following way:

$$a_2 \cdot b_2 = 2, \qquad a_1 \cdot a_2 \cdot b_1 = 1, \qquad b_1 = 1. \tag{3.4}$$

In a single-loop ADC the first integrator is the most sensitive building block for the overall performance [54, 56]. Due to this fact it makes sense to use the degree of freedom to permit a relaxed design-specification for the first integrator, that means to choose for the amplification factor a_1 a value smaller than one. The coefficient a_1 was set to $\frac{1}{2}$. Doing this reduces the slew-rate requirements of the OpAmp, which is directly related to lower power consumption. Another positive side effect is the damping of the noise towards the output contributed by the first integrator. The gain factor a_2 becomes bigger than one, in particular $a_2 = 2$, but this is not an issue, because any errors caused by the second integrator are first-order noise-shaped. Allowing the possibility of a canonic structure for the internal DACs both feedback coefficients b_1 and b_2 were fixed to one.

Figure 3.1, S_1, S_2 indicate how the integrators thermal noise is taken into account in the system simulations. These factors represent the kT/C-noise and the input reflected OpAmp-noise of the first and the second integrator. A linear model is used for the quantizer consisting of a constant gain factor g plus a white-noise source $E(z)$. The digital output value $Y(z)$ is fed back through two Digital-to-Analog Converter (DAC) in which the signal is scaled by the feedback coefficients b_1 and b_2. The noise contributed by the reference voltage circuit (S_3) is the dominant source in the feedback path. The jitter of a provided clock source imposes additional noise occurring at the input and at the reference voltage where analog signals are sampled. Superposition of all the mentioned noise sources defines the achievable dynamic range (DR) of the delta-sigma modulator. Each noise source with its specific transfer function towards the output was modeled within a MATHCAD platform [18]. This platform covers all relevant noise sources and reveals the dominant noise contributors. It is worth mentioning that for the broadband OpAmp- and reference-noise the aliasing mechanism applies.

Figure 3.2 shows the noise contributors at the output of the modulator in the frequency band up to 300 kHz. (0 dBFS refers to a sine-wave with 1.8 Vpp.) The dashed line marks the 2nd-order noise-shaped quantization noise. The dotted line

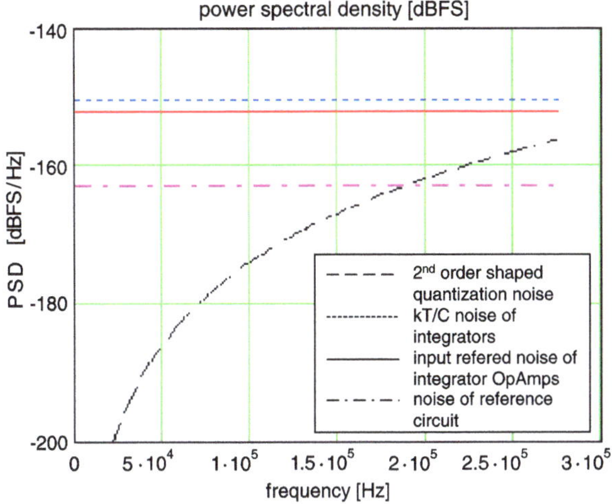

Fig. 3.2 Noise contribution, output referred

Table 3.1 Capacitor-values of low-power $\Delta\Sigma$-modulator

	1st integrator	2nd integrator
Sampling capacitor	1.05 pF	350 fF
Integrating capacitor	2.1 pF	175 fF
Feedback D/A capacitor	$7 \cdot 150$ fF	$7 \cdot 50$ fF

shows the added kT/C-noise floor of both integrators. The impact on the input referred noise of the integrator's OpAmps is depicted with a solid line. The noise originated in the reference circuit is shown as a dash-dotted line. As illustrated in Fig. 3.2 the specification for noise contributors are balanced in such a way, that the capacitor values can be chosen relatively small. Targeting to decrease the OpAmp's effective load capacitance, the values for the sampling capacitor of first and second integrator were carefully chosen. While the first-integrator capacitor is limited by noise considerations, the second-integrator capacitor is limited by design-constraints only. Deep sub-micron technologies suffer from high flicker noise being dominant up to several MHz. Choosing OpAmps as the 2nd dominant noise contributor helps to lower g_m in order to do a low-power design. The capacitor ratios of the SC-integrator are defined by the filter coefficients, since the gain is equal to the ratio of sampling capacitor over the integrating capacitor ($\frac{C_S}{C_I}$). The resulting capacitor values are given in Table 3.1.

3.2 Circuit-Design

Mixed-signal design becomes quite challenging at low supply voltages. A major concern is the implementation of switches. Clock-boosting techniques are a possible solution [2] which were not used in this design. The provided technology makes it possible to use 0.18 μm regular devices and 0.45 μm dual-GOX MOS-devices running at 1.8 V and 3.3 V respectively. The target product demands a dual-gate-oxide process option due to interface requirements. Therefore it was reasonable to use these devices for switches causing no extra costs. Closer investigations revealed that OpAmps using regular devices are more power efficient than OpAmps built with 3.3 V devices for achieving the required speed-performance in terms of GBW. All analog modules were implemented fully differentially in Switched-Capacitor (SC) technique.

3.2.1 Clocking-Scheme

A two-phase non-overlapping clock is used to drive the circuit, similar to [4]. *PHASE1* and *PHASE2* denote the sampling- and the integrating-phase, respectively. During *PHASE1* the 1st integrator has its integrating phase, the 2nd integrator has its sampling phase. During *PHASE2* it is vice versa. As discussed in previous section, the 2nd integrator can be downscaled due to a relaxed noise specification. In this way the sampling capacitor can be reduced so that it represents a negligible increase of the load for the OpAmp in the 1st integrator. The internal SC-quantizer samples the output voltage during *PHASE2* and takes a decision during *PHASE1*. Latching of quantized data is required for the first integrator.

Generally, SC integrators do not require high dynamic performances during the sampling-phase. This job is shifted to any previous amplifier that drives the sampling capacitor. Exploiting this property, dynamic-biasing of the amplifier can be implemented. As depicted in Fig. 3.3 the overall timing of the delta-sigma modulator was adapted in order to permit the introduction of a *dynamic-biasing technique*. Φ_1 and Φ_2 denote the clock-signals for the odd and even clock-phase, respectively. Φ_{1D} and Φ_{2D} are the delayed clock-signals to avoid clock feedthrough and charge injection during switching transitions [72]. Conventionally, both integrators perform the sampling and integrating during the same half-clock phase. The proposed solution swaps the timing in such a way that while the 1st integrator integrates, the 2nd one samples and vice versa. In our approach the bias-current of the output transistor stage is switched between two values with the rhythm of the clock frequency. This solution offers the possibility to reduce the rms-value of the integrator current of about 30 %. A further benefit is the more balanced capacitive loading of the reference voltage buffer, which relaxes the block specification for the on-chip reference circuit.

3.2.2 Switched-Capacitor Integrator

The implemented integrator is depicted in Fig. 3.4 including the sampling capacitors and the fully differential 3 bit DAC. The switches are implemented with single dual-Gate-Oxide (GOX) nMOS devices using a nominal gate voltage of 3.3 V. The sizes

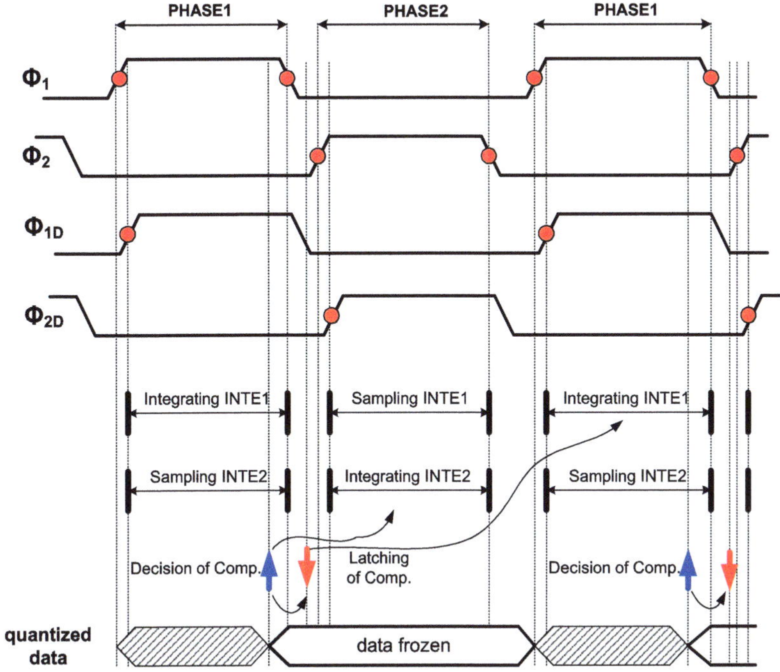

Fig. 3.3 Overall timing-diagram

of the switches were designed in such a way that for small signals the on-resistance is in the range of 300 Ω. That on-resistance is varied by bigger signal-amplitudes within an acceptable small span due to the high overdrive voltage. This choice provides a sufficiently small RC-time-constant resulting in a fast settling transient. The sampling-capacitors and DAC-capacitors were arranged in such a manner that there was a constant charge drawn from the references in each clock-cycle. A different arrangement, also used in [64], allows a reduction in the capacitor size by a factor of two at the cost of a signal dependent charge being drawn from the references. Essentially, we have chosen the more robust implementation at the cost of increased power consumption. In addition, the gain-factor b_1 (see Fig. 3.1) can be chosen to be non-unity to scale the input range of the modulator.

Multi-bit delta-sigma modulators are known to have one major drawback, which is their sensitivity to the internal DAC inaccuracies. Mismatch between the DAC elements can generate harmonic distortion. Basically, the internal DAC needs to perform as well as the overall modulator. A solution for this problem can be found in Dynamic Element Matching (DEM) techniques, which basically convert internal DAC element errors to high frequency noise. Thereby highly linear oversampling DACs can be built with only moderate matching requirements for the DAC elements. A bibliography of such techniques is given in [65]. Those techniques are being developed since 1988 starting with randomization of the DAC-elements [32]. The methods are continuously improved with respect to implementation efficiency

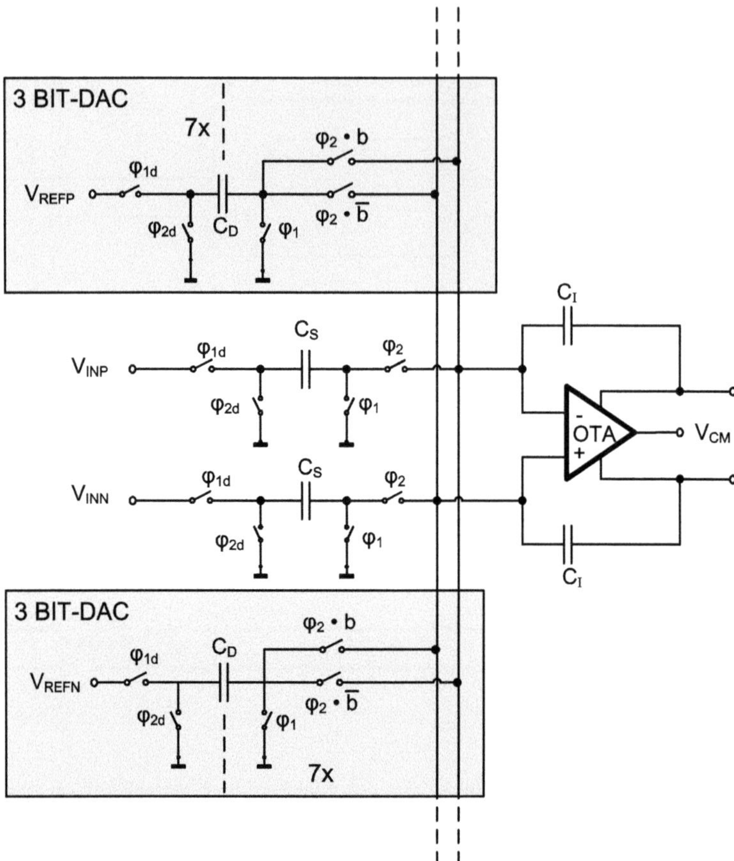

Fig. 3.4 Switched-capacitor integrator and DAC

and order of shaping. Since the presentation of the Baird and Fiez paper [46] in 1995 and the disclosure of the ADC design of Brooks et. al. [64] in 1997, these techniques are well established in the delta-sigma design community, allowing efficient and robust implementations of delta-sigma ADCs with resolutions of more than 14 bits and bandwidths beyond 1 MHz [4, 17, 27, 77].

The disadvantage of the DEM techniques is that they introduce an additional delay inside the modulator's feedback loop, requiring other components to operate faster. We could avoid to use DEM techniques in our design, because sufficiently good element-matching could be achieved by simple layout measures without any drawbacks. Simulations revealed that a standard deviation of the 3 bit DAC-element mismatch smaller than 0.05 % is sufficient not to degrade the modulators SNR performance significantly and to achieve harmonic distortion below -100 dB full-scale. The 3 bit DAC is implemented in switched-capacitor (SC) technique, utilizing seven unity-capacitors (see Fig. 3.4). The absolute value of the capacitors determines the matching properties as well as the kT/C-noise introduced into the

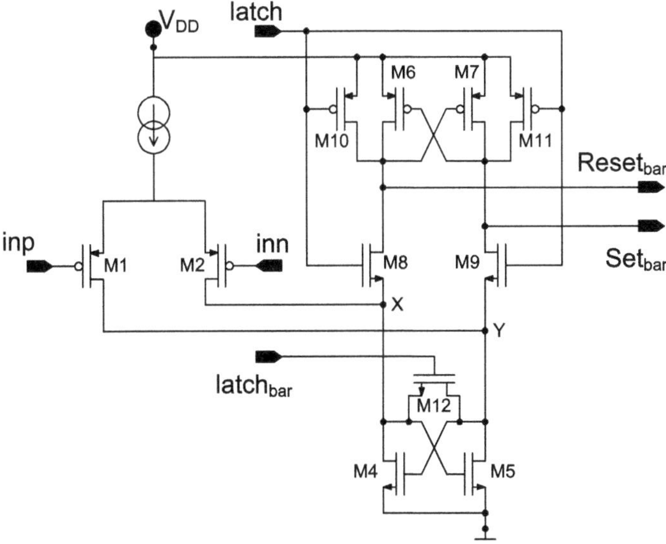

Fig. 3.5   Schematic of latch-circuit

circuit. It was found that the kT/C-noise constraint requires larger capacitors than
the element-matching parameters would require for the used MIM (metal-insulator-
metal) capacitors of the 0.18 μm CMOS process. The finally chosen capacitance
values are shown in Table 3.1.

3.2.3 Quantizer

The internally used 3 bit flash-ADC is implemented fully-differentially in SC-
technique employing a single-stage comparator as depicted in Fig. 3.5 and described
in [14].

Within the operating band, the noise contribution of the noise shaped 3 bit ADC is
very small compared to other noise sources (see Fig. 3.2). Due to this, the quantizer
is not a limiting factor concerning noise-performance. Mismatch and offset are no
issues for the design also the quantizer does not represent a critical load for the
2nd integrator. Thus, during implementation of the quantizer, the focus was put on
design robustness.

3.2.4 Reference Buffer

Choosing a value for the reference voltage is a trade-off between maximizing the
dynamic-range and achieving sufficient linearity. The derived coefficients cause an

integrator-output voltage-swing 20 % bigger than the reference voltage V_{REF}. Generally, the maximum realizable reference voltage is limited by the possible signal swing of an OpAmp and can be estimated as follows:

$$V_{REF} < \frac{V_{DD} \cdot 0.95 - 2 \cdot V_{Dsat}}{1.2}. \tag{3.5}$$

Where the factor 0.95 stands for supply variations and the factor 1.2 for the 20 % increased signal-swing at the integrator output. Considering some additional margin the reference voltage was chosen to be 0.9 V (differential). This choice was proven to be a reasonable trade-off between sufficient linearity (low V_{REF}) and high dynamic range (high V_{REF}). Miller OpAmps serve as reference buffers in this design. The power drain of these elements is about 40 % of the total power consumption. Although we investigated alternatives to improve the power efficiency, we have chosen our approach for reuse reasons.

Instead of having class-A buffers directly driving the capacitors, a class-AB buffer can be used. Simulations show that the references would draw only 25 % of the total current. A different approach to reduce reference buffer current drain is presented in [27], where the transient charge is delivered from an external capacitor having its voltage controlled by a class-A OpAmp. Using the latter approach power dissipation could be further reduced at the cost of using external components and increased package pin-count.

3.2.5 Operational-Amplifier

Generally, an OpAmp topology with high output swing is required for a power-efficient implementation. In larger feature size technologies telescopic- or folded-cascode-structures were used for switched-capacitor circuits [56, 71, 72]. In deep sub-μ technologies, these single stages OpAmps provide poor DC-gain due to high g_{DS}. In addition, they consume at least $4 \times V_{Dsat}$ of headroom. Small feature-size technologies require several cascaded transistor stages to provide sufficient DC-gain and high output voltage-swing. A two-stage Miller-OpAmp [71] approach offers sufficient DC-gain and reduces the swing just by $2 \times V_{Dsat}$. Cascading of gain stages generally consumes more current at the benefits of an improved DC-gain and a higher output swing. In spite of this fact, a lowered power supply voltage makes it possible to reach competitive power consumption. A detailed analysis revealed that the converter's power efficiency (PE) using a Miller OpAmp is better than the PE employing single-stages OpAmps if the supply voltage is below 2 Volts along with high output swing requirements. A two stage Miller OpAmp was chosen to meet the building block specifications.

These specifications were defined as follows: To overcome finite-gain effects the open-loop gain was chosen to be bigger than 55 dB. The finite closed-loop pole frequency must be higher than three times the clock frequency to not compromise the desired SNR. Considering a clock-frequency of about 53 MHz and a reference voltage of 0.9 Volts leads to a required lower limit for the slew-rate of 350 $\frac{V}{\mu s}$. But this comes for free because fulfilling the noise- and speed-requirements already result in an higher slew-rate of about 500 $\frac{V}{\mu s}$. The equivalent white-noise density over the band of interest referred to the input were set to be smaller than 9 $\frac{nV}{\sqrt{Hz}}$. The OpAmp of the first integrator results in an open-loop GBW of 400 MHz driving an effective load of 4.8 pF. Maintaining a phase-margin of better than 70° requires a compensation-capacitance of 750 fF for proper pole-splitting between the dominant- and non-dominant pole of the Miller-OpAmp.

The circuit of a two-stage fully-differential amplifier (Fig. 3.6) includes a differential input stage, two output stages and a common-mode feedback circuit coupled to the output nodes of the amplifier. The operational amplifier (OpAmp) is also provided with a *control voltage* (Vc) which controls the common-mode voltage at the output nodes (OUTP, OUTN). The control voltage (Vc) is applied at the gate of the active load (Mla and Mlb) of the input differential-pair to set the common-mode half-way between the voltages of the supply rails (VDD and VSS). In single-stage fully differential amplifiers such as folded-cascode amplifiers, the SC common-mode feedback circuit may be implemented extremely simple since it does not require any additional circuitry [56, 71, 72].

The main requirements for the common-mode feedback (CMFB) control-loop are the inherent stability and no degradation of the output voltage headroom, which becomes severe in deep sub-micron technologies. In the case of a two-stage OpAmp the CMFB-circuit cannot be realized the same as for a single-stage amplifier according to [56, 71, 72], because two inversions would occur between the control voltage (Vc) and both outputs (OUTP, OUTN). This would result in positive feedback that would make the system unstable. Achieving stability one additional inversion has to be done with a PMOS current-source consisting of transistor Mb3 and Mb4. The transistor Mb3 provides a constant part of the control current (Ic), whereby a variable part of current Ic is generated by the transistor Mb4. The gate-source voltage of Mb4 is derived by the mean-value of both output voltages (Vm). The current partitioning into a constant- and variable- portion helps to stabilize the common-mode loop and avoids any start up troubles. The diode-connected NMOS transistor Mb10 converts the control current (Ic) into a control voltage (Vc) for biasing the active load (Mla and Mlb) in a proper manner. The possible amplifier's output voltage of that solution is not affected by the proposed CMFB-circuit, since only two capacitors are connected to the outputs. All switches inside the switched-capacitor CMFB-circuit are implemented with dual-GOX devices biased with a clock derived from the higher supply voltage (3.3 Volts).

Due to the requirements of SC-circuits, the step-response becomes quite important. Therefore the closed-loop phase-margin has to be bigger than 76-degrees to get a critical damped step response. The phase-margin is approximately proportional to the ratio of $\frac{g_{m2}}{C_{load}}$, where g_{m2} denotes the transconductance of the 2nd transistor stage (M2a and M2b). The gain-bandwidth-product is approximately proportional to the ratio of $\frac{g_{m1}}{C_C}$, where g_{m1} denotes the transconductance of the input stage (M1a and M1b). Taking this as well as the noise and slew-rate considerations into account

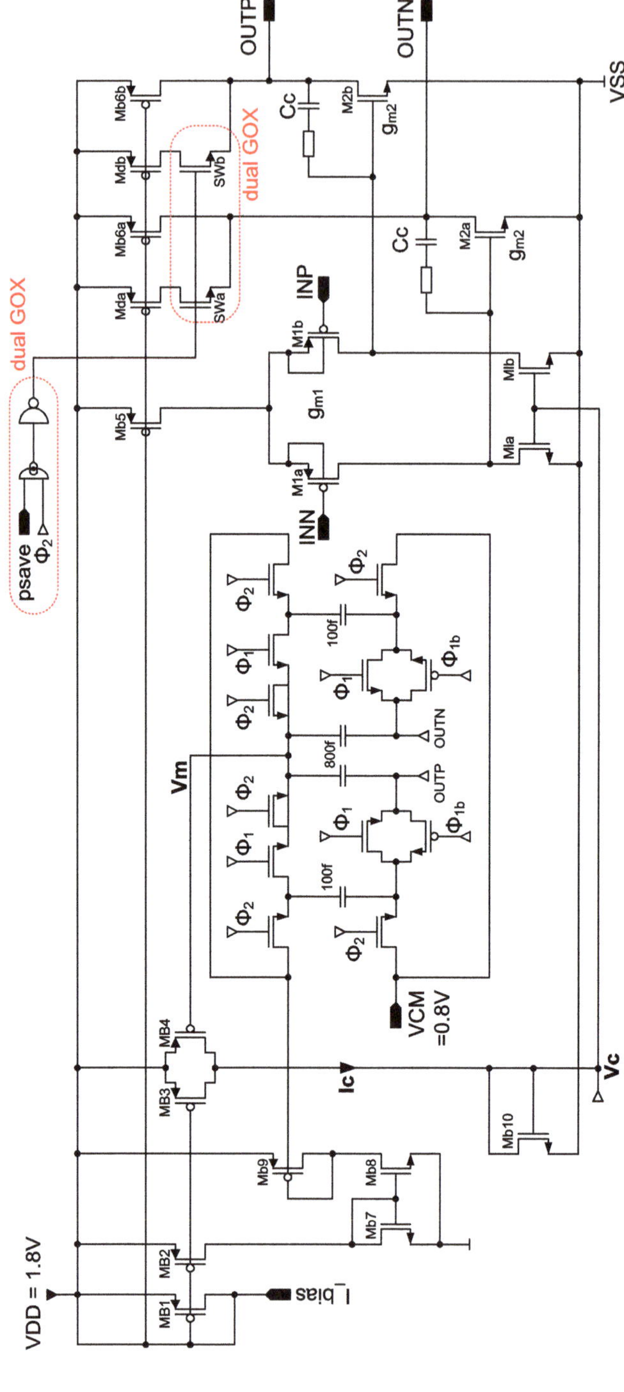

Fig. 3.6 Schematic of operational amplifier

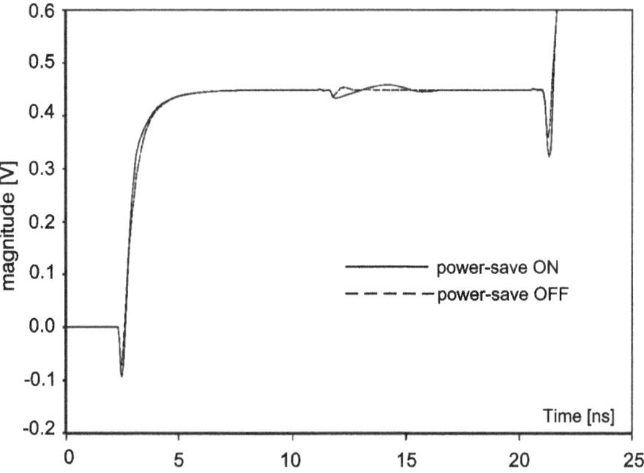

Fig. 3.7 Settling transient of first integrator

the drain-current in the output-stage is more than four times bigger as compared to the drain-current of the input differential pair. This fact clearly indicates that power saving can only be done efficiently in the output stage. The chosen approach was to control the bias-current dynamically. In general, switched-capacitor integrators comprise two phases; the so-called sampling phase and integrating phase. During the sampling phase, the sampling capacitor is driven by a preceding amplifier and the charge stored on the integrating capacitor is not changed. During the integrating phase the integrator's OpAmp performs the charge redistribution and shifts the stored charge from the sampling capacitor onto the integrating capacitor. Furthermore, the OpAmp has to drive the capacitors of the next stage. In other words, the OpAmp is utilized alternately in a heavy and easy manner. During integrator's integration phase, the OpAmp has to deliver a huge current. During its sampling phase, the output current of the OpAmp is heavily decreased. This fact makes it possible to introduce a *dynamic-biasing* technique. Doing this, the OpAmp design has to be done twice according to both phases. During the sampling phase the closed loop of the OpAmp has to be kept stable ignoring any slew-rate and gain-bandwidth constraints. The closed-loop phase margin under worst-case conditions can be lowered to 30 degrees. This relaxed specification has been employed to reduce the drain current of the output stage (M2a and M2b) by 70 %. More details about the dynamic-biasing technique can be found in [47].

Keeping the circuit as simple as possible, the biasing for each output transistor consists of two current sources: Devices Mda, Mb6a for the output transistor M2a and devices Mdb, Mb6b for M2b, respectively (see Fig. 3.6). The current sources Mb6a and Mb6b provide continuously flowing DC-current. The additionally needed dynamic portion of the drain current is provided by the current sources Mda and Mdb. The additional current is driven with the rhythm of the clock-signal Φ_2. The sum of both currents delivers the necessary output-current for achieving sufficient settling accuracy. Exploiting this *dynamic-biasing* technique offers a reduction of the overall power consumption in the range of 30 %. This described mode is called *power-save mode*. Figure 3.7 shows the settling behavior for disabled and enabled power-save mode. The integration process happens during the phase from 3 to 10 ns, from 10 to 20 ns the sampling phase takes place. No significant difference can be seen between both curves, which demonstrates the robustness of that proposed power-save technique [47].

3.3 Measurement Results

The modulator was implemented in a 0.18 μm 4-metal CMOS technology using a dual-gate-oxide process option for the switches. The linear capacitors were implemented as metal-insulator-metal (MIM) devices. The evaluation of the prototype needs to be done very carefully due to practical constraints in measurement [47]. The chip-size results in 1.9×1.9 mm^2, which is relatively small for a standard CQFP-64 package resulting in bondwire length of more than 7 mm. Such long bondwires exhibit an inductance bigger than 5 nH forming a parasitic LC-tank together with the SC-input stage of the ADC. Avoiding undesirable effects involves a continuous-time pre-filter stage for signal decoupling reasons. A fully differential voltage generator is applied to the on-chip pre-filter which drives the delta-sigma ADC. The 2nd order Butterworth-filter has a -3 dB corner-frequency of 500 kHz providing an anti-aliasing filter. The on-chip references are implemented fully differentially in a feed-back configuration. The quantizer uses a separated reference circuit. The overall bias-currents and reference voltages are derived from an on-chip programmable biasing circuit including a bandgap. Figure 3.8 shows the complete layout of the test-chip. A photograph of the die is shown in Fig. 3.9. The ADC core-area results in 0.21 mm^2, all required reference buffers consume additionally 0.18 mm^2 of chip area. The whole signal-path is implemented fully differential. Avoiding crosstalk issues requires a separated supply for the analog and the digital part as well as the central biasing .

The quantizer's data output stream is connected to digital output pads. Measurement data were obtained utilizing a logic analyzer. Frames of 32k samples were captured. Post-processing was performed off-line using MATLAB. The clock-source (53 MHz) and the signal-source (100 kHz) are synchronized via a proper synchronization interface. For an accurate analysis of the inband-spectrum (0–276 kHz) the following postprocessing actions are done:

- Filtering the out-of-band noise to prevent the aliased out-of-band noise from leaking into the band of interest.
- Skipping the signal samples on the beginning of the signal vector to prevent leakage coming from filter settling and cutting the vector to an integer number of signal periods.
- Using a window with good roll-off characteristic to reduce remaining leakage, a hanning window was chosen.

The Power Spectral Density (PSD) plot of a Fast Fourier Transform (FFT) result shows statistical variations of the bins, caused by random noise superimposed to a

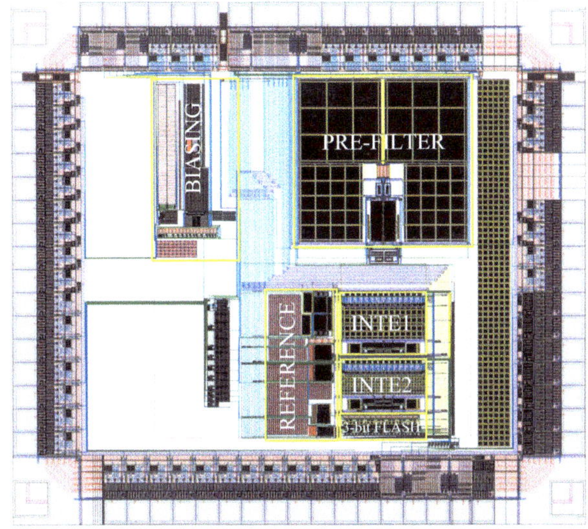

Fig. 3.8 Layout of the 0.18 μm test-chip

Fig. 3.9 Photograh of the 0.18 μm test-chip

signal. Thus, measurement of discrete tones with power P_s is influenced by noise. For the relative error of the signal power estimator $\hat{P}_s$ one can easily derive

$$\frac{\sigma(\hat{P}_s)}{P_s} \approx 2 \cdot \sqrt{\frac{N_{fl}}{P_s}} \qquad (3.6)$$

where N_{fl} is the noise-floor level of the PSD plot and is equal to the noise power P_n divided by the number of samples N fed to the FFT. This means that decreasing signal amplitude causes an increasing error of the signal power estimator $\hat{P}_s$. If the

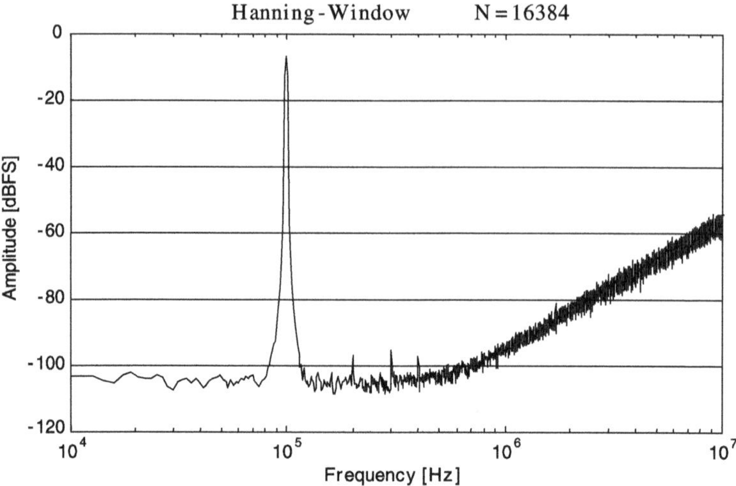

Fig. 3.10 Output data-stream PSD plot (-7 dBFS)

signal power P_s is equal to the noise-floor level N_{fl} the error of the measured signal power is equal to the N_{fl}. It is worth mentioning that N_{fl} represents the noise power per bin and can be reduced by increasing the number of FFT-points N. Nevertheless, the physical noise expressed in $V/\sqrt{Hz}$ remains unchanged.

For improvement of the measurement-accuracy in terms of noise and distortion, spectral averaging is introduced. For harmonic distortion components with amplitudes close to the noise floor level this technique is very effective. For this method a N-point FFT-spectrum is recorded M-times. Then the average PSD is calculated. Other postprocessing (SNR, THD, etc.) remains unchanged. For the relative error of the signal power estimator $\hat{P}_s$ follows

$$\frac{\sigma(\hat{P}_s)}{P_s} \approx 2 \cdot \sqrt{\frac{\frac{P_n}{M \cdot N}}{P_s}}. \tag{3.7}$$

For M captured signal frames of length N the statistical error of the determined signal amplitude decreases by $\sqrt{M}$. Another way of getting more accurate spectral analysis is to capture more signal-samples, that means to get more points N for the FFT lowering the N_{fl}. With the available measurement setup, it was not possible to save more than 32k samples.

The FFT result for a -7 dBFS input signal is depicted in Fig. 3.10 in which the spectrum was averaged 10-times as discussed above. (0 dBFS refers to a sine-wave with 1.8 Vpp.) The band of interest is dominated by white-noise. Distortion attracts attention with -90 dBc for the 2nd harmonic and -88 dBc for the 3rd harmonic. At 400 kHz, the second-order shaped-noise becomes dominant.

Figure 3.11 shows the measured Signal-to-Noise Ratio (SNR) and Signal-to-Noise and Distortion Ratio (SNDR) versus the normalized input signal $\frac{V_{in}}{V_{REF}}$. Up to -6 dB the noise-energy is dominant over the energy of the harmonics.

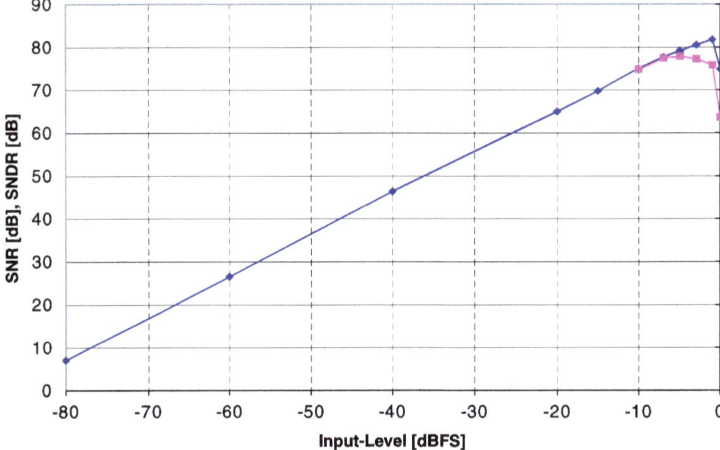

Fig. 3.11   Measured signal-to-noise and distortion ratio

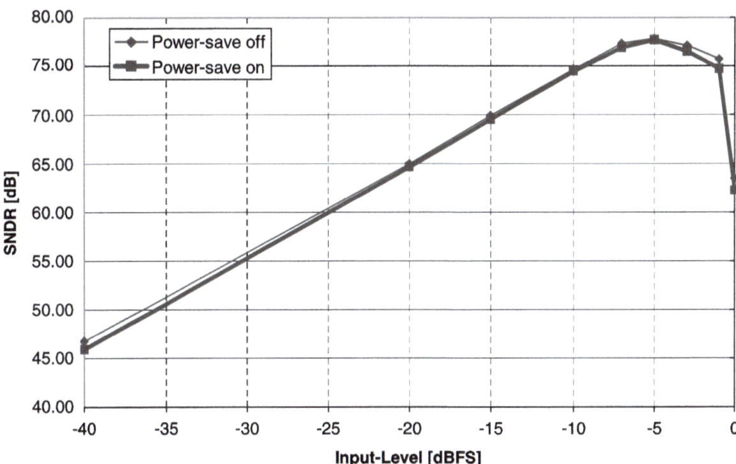

Fig. 3.12   Impact of power-save mode on performance

Figure 3.12 shows the SNDR of the ADC with and without both integrators in power-save mode, as previously discussed. The current-consumption of the ADC core drops from 5.1 mA to 3.6 mA.

Reducing the overall bias-current of the integrators by 25 % results in lower performance (Fig. 3.13). Additionally enabling the power-save mode, the performance is lowered again. The SNDR degradation is caused by increased harmonic distortion. Especially at higher input levels, distortion increases by up to 8 dB, if the overall biasing gets reduced down to 75 % and the dynamic-biasing technique is enabled as well.

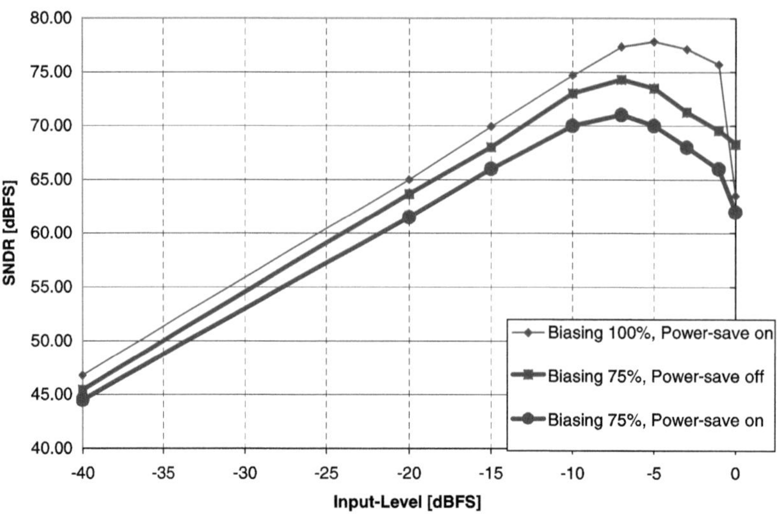

Fig. 3.13 Influence of integrator bias-current on performance

Table 3.2 Measured performance summary of 0.18 μm A/D-converter

Signal bandwidth	276 kHz
Sampling frequency	53 MHz
Oversampling ratio (OSR)	96
Reference voltage	0.9 V (diff.)
Dynamic range (DR)	85 dB
Peak SNR	82 dB
Peak SNDR	78 dB
Power ADC core (1.8 V/3.3 V)	9 mW/6 mW
Power reference (1.8 V)	10 mW
Pre-filter (1.8 V)	9 mW
Technology	0.18 μm 1.8 V CMOS 4ML
FOM (DR) (ADC-core I ADC + Ref)	1.8 pJ/conv I 3.0 pJ/conv
FOM (peak SNR) (ADC-core I ADC + Ref)	2.6 pJ/conv I 4.3 pJ/conv
FOM (peak SNDR) (ADC-core I ADC + Ref)	4.1 pJ/conv I 6.8 pJ/conv

The layout plot in Fig. 3.8 demonstrates that this presented design-strategy results
in an attractive silicon area compared to similar delta-sigma converter [4]. A significant part of the silicon area needs to be spent for power routing. Clearly, the area
is dominated by capacitors whereas transistors play a minor importance. 15 % area
is used by the internal 3 bit quantizer. Table 3.2 summarizes some of the important measurement data. The dynamic range of the measured samples spreads up to
88 dB.

3.4 Conclusion

A low voltage SC delta-sigma modulator has been implemented in a 0.18 μm technology [47]. The architecture was chosen to employ the high-speed core devices in 0.18 μm CMOS enabling an Over-Sampling Ratio (OSR) of 96. Thus, it was possible to go for a single-loop modulator with 2nd order noise shaping leading to an attractive silicon area and power consumption. The FOM (2.7) results in 1.8 $\frac{pJ}{conv}$ and 3.0 $\frac{pJ}{conv}$ taking into account the power drain of the ADC and the ADC including all reference buffers, respectively. Keeping the capacitor values small requires OpAmps with a high output swing capability to not compromise the Dynamic Range (DR). Using a Miller-OpAmp gives a better power-efficiency as compared to single-stage OpAmps for the chosen architecture. An innovative dynamic-biasing circuit is introduced for lowering the power consumption of the SC-integrators at an expense of increased complexity. 14 bit performance was achieved without dynamic element-matching algorithms, due to the excellent metal-insulator-metal capacitor matching of the deep-sub-micron process. As compared to other designs with similar requirements built in larger feature-size technologies [4], the power efficiency was not improved significantly, but the silicon area could be drastically reduced.

In this design, the ADC-core was the main target for improvement. The main purpose of the test-chip was to demonstrate that a single-loop modulator is suitable as an architecture to fulfill all ADSL requirements, because in the past, MASH structures had been used [3, 4]. There would be additional room for a power drain reduction by modifying the outer core-blocks. Especially the 3.3 V digital part for clock generation and switching consumes at least 67 % compared to the power drain of the whole ADC-core. This could be improved by reducing the digital supply down to 1.8 V and replacing the 3.3 V switch-drivers by level-shifters.

Chapter 4
A Feed-Forward Delta-Sigma Converter for ADSL

The growing market for broadband access services is the driving force for Digital Subscriber Line (DSL) applications. As illustrated in Fig. 1.1, DSL allows for simultaneous transmission of digital data and the Plain-Old Telephone Service (POTS) signal on a single copper-pair. Among several DSL approaches the Asymmetric Digital Subscriber Line (ADSL) technology was standardized. Generally, ADSL can transport more than 8 Mbit/s from the Central Office (CO) to the customer (downstream) and more than 1 Mbit/s upstream. As shown in Table 1.1, several options were standardized such as ADSL+ to offer a higher bit-rate for downstream at the cost of reduced subscriber loop-length. Furthermore, line-testing features were introduced for performance monitoring requiring an analog bandwidth up to 1.1 MHz for the A/D-converter.

A power- and area-optimized multi-bit delta-sigma ADC including a reference-voltage buffer is introduced [48]. The proposed A/D-converter is targeted for an ADSL Central-Office (CO) application implemented in a 0.13 μm CMOS technology. A new feed-forward topology is developed permitting the usage of power-efficient single-stage amplifiers. All blocks are composed of regular threshold voltage devices biased from a single 1.5 V supply only. An area- and power-efficient realization of a 2nd-order, single-loop, 3 bit modulator with high oversampling-ratio (OSR = 192) is presented. The delta-sigma modulator features a 14 bit and 13 bit dynamic-range over a 276 kHz and 1.1 MHz signal bandwidth, respectively. The measured power consumption of the ADC-core is 8 mW only. Including an on-chip reference buffer the total power consumption results in 15 mW. The FOM of the ADC-core as described in (2.7) results in 0.7 $\frac{pJ}{conv}$.

The AFE integration evolution tries to integrate more and more lines per board. The dissipated power is one limiting factor for high integration apart from any cross-talk problems [42]. Therefore, in ADSL CO applications the power consumption is one of the most important criteria. This has a strong impact on the architectural considerations.

R. Gaggl, *Delta-Sigma A/D-Converters*, Springer Series in Advanced Microelectronics 39, 75
DOI 10.1007/978-3-642-34543-2_4, © Springer-Verlag Berlin Heidelberg 2013

4.1 Architectural Considerations

The developed Analog-to-Digital Converter (ADC) targets a CO line-card application [42]. Achieving a low-power converter requires careful considerations on system-level design [48]. During the last years it turned out that delta-sigma modulators are preferentially used compared to pipelined converters in the field of ADSL [3]. In principle, delta-sigma converters can be divided into single-loop structures and Multi-stAge noise-SHaping (MASH) architectures, as described in chapter two. The MASH architecture uses low-order modulators in each stage to make the system stable and a digital cancellation logic is used to cancel quantization noise. In general, mismatches between each stage and the digital cancellation circuit my cause a performance degradation as shown in condition (2.20). In other words, the performance of a MASH is quite sensitive to circuit non-idealities, such as nonzero switch resistances (2.30), (2.28) and finite DC-gain as well as the limited bandwidth of the OTA (2.24). Since it is hard to design low-power OTAs with quite high DC-gain in deep-submicron technologies, MASH topologies were not considered adequate and a single-loop modulator was selected.

Due to the stability constraints of high-order single-loop topologies [54, 56], the order of noise-shaping was chosen to be two. This approach needs a small number of building blocks, hence the necessary silicon area becomes quite small. As a trade-off, the OSR was chosen to be 192, resulting in a sampling frequency of 105 MHz. To bring the quantization noise floor well below the 14 bit level, a 3 bit internal quantizer was chosen. It is worth mentioning that the high OSR makes it also possible to achieve a SNR of 13 bit over a signal bandwidth of 1.1 MHz for supporting line-testing features. MOS devices in deep-submicron technologies feature high-speed at the cost of fairly high output-conductance (g_o) resulting in low DC-gain. Therefore, this choice is very much in favor with the properties of deep-submicron technologies. The loopfilter is implemented in Switched-Capacitor (SC) technique employing the good matching properties of integrated capacitors resulting in accurate filter coefficients. Furthermore, SC modulators are less sensitive to clock jitter as compared to CT modulators, as already discussed in the second chapter. This fact is crucial for System on Chip (SoC) solutions in which the very low jitter clock-generation and clock-distribution between several channels might become power dominant.

4.1.1 A Feed-Forward Approach

A block diagram of a conventional 2nd-order single-loop delta-sigma modulator was already shown in Fig. 3.1. System level simulations with sinusoidal input signals of the proposed conventional delta-sigma modulator show the following necessary signal swing at the integrators' output terminals (see Table 4.1).

The full-scale value is defined by the provided reference voltage V_{ref}. The needed output swings of both integrators must be larger than V_{ref} to keep the occurring harmonic distortion below the specification. It is obvious that the choice of

Table 4.1 Max. signal-swing in a conventional $\Delta\Sigma$-modulator

Input signal	Swing of 1st integrator	Swing of 2nd integrator
-3 dB full-scale	-1.3 dB full-scale	0 dB full-scale
-6 dB full-scale	-3.8 dB full-scale	-0.7 dB full-scale

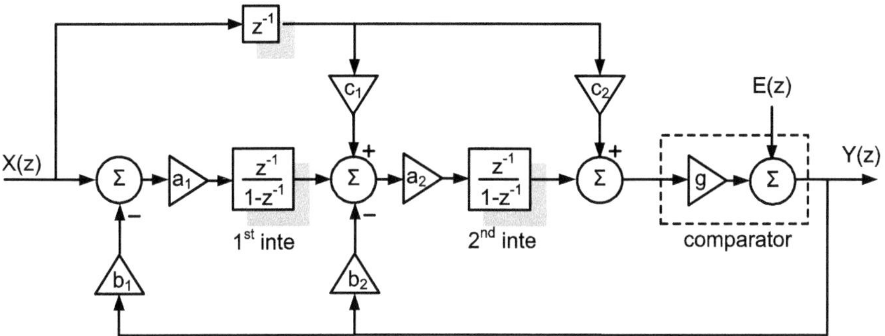

Fig. 4.1 2nd-order $\Delta\Sigma$ modulator with feed-forward approach

V_{ref} is absolute proportional to the dynamic range. That means, a trade-off has to be accepted between high SNR (large V_{ref}) and low distortion (small V_{ref}). This fact makes low-voltage designs even harder due to reduced possible signal-swing. In order to improve that issue multi-stage OTAs can be used to provide high output swing [71, 72]. But each transistor stage needs a bias current to generate a certain transconductance g_m. To lower the power consumption of multi-stage OTAs dynamic biasing techniques could be used as introduced in chapter three. The dynamic biasing technique offers a reduction in power drain at the expense of increased circuit complexity.

It is more promising to address the low-power issue already on architectural-level as finally on circuit-level. The reason for the needed large signal-swing lies in the fact that the input signal propagates through both integrators to the output of the delta-sigma modulator. One solution to overcome this fact is to introduce a *feed-forward approach* of the input signal as depicted in Fig. 4.1. Both *feed-forward paths* c_1 and c_2 *bypass* the input signal so that inside the noise-shaping loop the signal itself vanishes. Only the shaped quantization noise signal determines the necessary integrator output swing. The proposed second-order modulator employs two delaying integrators which is desirable because it allows the OTAs in each integrator to settle independently of each other, thereby relaxing their speed requirements. The linear transfer function of a discrete-time delaying integrator [54] results in

$$H(z) = \frac{z^{-1}}{1 - z^{-1}}. \tag{4.1}$$

Presuming unity gain g for the multi-bit quantizer, linear analysis of modulator depicted in Fig. 4.1 shows that the Signal Transfer Function (STF) is

$$STF(z) = \frac{c_2 \cdot z^2 + (a_2c_1 + a_2a_1 - 2c_2) \cdot z + c_2 - a_2c_1}{z^3 + (a_2b_2 - 2) \cdot z^2 + (a_1a_2b_1 - a_2b_2 + 1) \cdot z} \qquad (4.2)$$

and the Noise Transfer Function (NTF) results in

$$NTF(z) = \frac{(1 - z^{-1})^2}{(a_1a_2b_1 - a_2b_2 + 1) \cdot z^{-2} + (a_2b_2 - 2) \cdot z^{-1} + 1}. \qquad (4.3)$$

In single-stage loops, the first integrator is the most sensitive building block for the overall performance [54, 56]. The choice of the modulator's coefficients was constraint in such a way that the zeros of the noise transfer function are placed at DC and that the first integrator receives a relaxed specification. That means, the coefficients a_1 and a_2 were set to $\frac{1}{2}$ and 2, respectively. Allowing the possibility of a canonic structure for the internal DACs both feedback coefficients b_1 and b_2 were fixed to one. This solution is quite similar to the modulator presented in the third chapter except for the increased OSR and the feed-forward of the input signal. In order to achieve $NTF(z) = (1 - z^{-1})^2$ constrains the choice for coefficients in the following way:

$$a_2 \cdot b_2 = 2, \qquad a_1 \cdot a_2 \cdot b_1 = 1, \qquad b_1 = 1. \qquad (4.4)$$

Choosing the coefficients c_1 and c_2 in such a way that on one hand the signal-swing of the integrator's output ports becomes very small and on the other hand the STF of the proposed delta-sigma modulator is at least in the band of interest equal to unity results in the value one for c_1 and c_2:

$$c_1 = 1, \qquad c_2 = 1. \qquad (4.5)$$

The resulting NTF is equal to the desired function of $NTF(z) = (1 - z^{-1})^2$ and shapes the quantization error by $+40$ dB per decade, as shown in Fig. 4.2. As a trade-off, the resulting STF exhibits two real zeros occurring sufficiently above the band of interest, as shown in Fig. 4.3. Additionally, the STF consists of three clock delays (z^{-3}) causing a phase-shift without affecting the magnitude. The position of these zeros and poles do not deteriorate the converter's performance. The peaking at 33 MHz goes up to $+14.5$ dB. Thus, the preceding anti-aliasing filter needs to take care for damping any out of band disturbers accordingly. Any stability risk caused by the feed-forward approach was strongly investigated in several behavioral-simulation runs. Different input-frequencies and -levels were applied to the delta-sigma modulator to prove the stability in an empirical manner.

The feed-forward approach reduces the required signal-swing of both integrator stages. The occurring swing for a -3 dB and -6 dB full-scale input signal is shown in Table 4.2. The input signal bypasses via c_1 and c_2 both integrator stages, thus, the output swing is determined mainly by the shaped quantization error. Therefore, the signal swing is almost independent of the input amplitude. The proposed solution permits an implementation of a power efficient single-stage OTA such as a folded-cascode OTA without hurting the distortion specification by limited signal-swing capability.

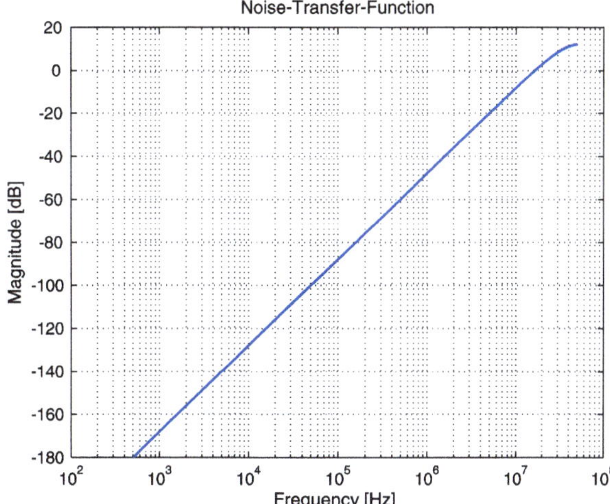

Fig. 4.2 2nd-order noise transfer function

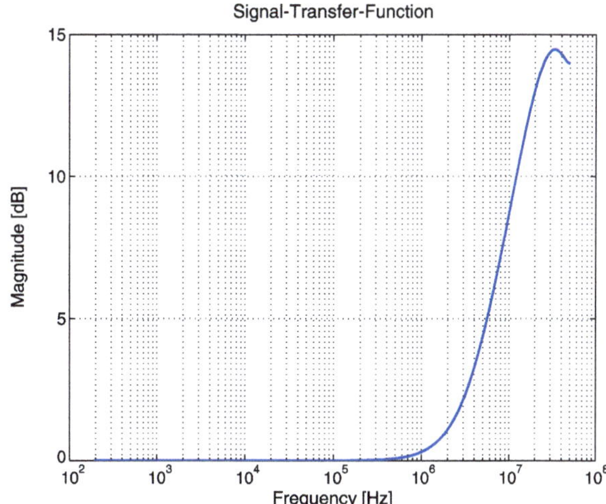

Fig. 4.3 Signal transfer function with a feed-forward approach

Table 4.2 Max. signal-swing in a $\Delta\Sigma$-modulator employing feed-forward

Input signal	Swing of 1st integrator	Swing of 2nd integrator
−3 dB full-scale	−16.9 dB full-scale	−7.3 dB full-scale
−6 dB full-scale	−16.9 dB full-scale	−7.3 dB full-scale

Each existing noise source inside the modulator with its specific transfer function towards the output was modeled within a MATHCAD platform as discussed in [18]. While the sampling capacitance of the first integrator is limited by kT/C-noise considerations, the second one is matching limited to achieve sufficient linearity by employing dynamic-element matching techniques. Linear capacitors are used

Table 4.3 Capacitor values of $\Delta\Sigma$-modulator with feed-forward approach

	1st integrator	2nd integrator
sampling capacitor	840 fF	280 fF
integrating capacitor	1.68 pF	140 fF

composed of standard metalization stack and thereby reducing the manufacturing complexity. Thus, as compared to MIM-capacitors, metal-sandwich capacitors cut costs per wafer at the expense of bigger mismatch between sandwich capacitors. Choosing OTAs as the 2nd-dominant noise contributor helps to lower g_m in order to do a low-power design. Thanks to the high OSR the absolute capacitance values could be kept small so that the effective load for both integrators and the preceding anti-aliasing filter allows a low-power design. The resulting capacitor values are given in Table 4.3. The capacitor ratios are defined by the filter coefficients, since the gain of the SC-integrator is equal to the ratio of sampling capacitor over the integrating capacitor ($\frac{C_S}{C_I}$). The impact on clock jitter was taken into account as well. The maximum tolerable long-term jitter can go up to $\sigma_{lt} = 150$ ps. As shown in the open literature, comparable CT-modulators require a clock jitter which is one order of magnitude smaller. This fact demonstrates the jitter robustness of SC-modulators against the CT counterparts.

4.2 Circuit-Design

The design of analog blocks in deep-submicron technologies is quite challenging. As the supply voltage goes down so does the available signal-swing. To reach a certain Dynamic Range (DR) the noise level must be lowered as well, possibly leading to an increase in power consumption and area. Delta-sigma modulators employing continuous-time circuitry have received a growing attention during the last years due to the possibility of achieving low-power consumption. Such modulators suffer strongly from circuit imperfections, as pointed out in the second chapter. Especially clock jitter limits the achievable resolution. Implementing an ultra-low jitter Phase-Locked-Loop (PLL) and a clock-distribution circuitry for a multi-channel AFE causes an increased power drain. To avoid this, a switched capacitor (SC) approach has been implemented. Such modulators are more robust with respect to circuit imperfections, making them very suitable for achieving the targeted performances.

The key parameter of a delta-sigma converter is the reference voltage V_{ref}. In [75] it is shown that the modulator power consumption is inversely proportional to the square of V_{ref}. Moreover, the upper limit V_{ref} is related to the output-swing capability of the OTA's. Therefore, maximization of the OTA's output swing is needed for low-power design. The feed-forward approach brakes the relation between the integrators output signal-swing and V_{ref}, giving a degree of freedom in choosing the OTA's output swing capability.

Table 4.4 Specification of first OTA (folded-cascode)

Gain-bandwidth-product	400 MHz
Open-loop gain at DC	>60 dB
Phase margin	$>70°$
Effective capacitive load	4.2 pF
Input referred noise-density	$2\,\text{nV}/\sqrt{\text{Hz}}$

The first problem to face in low voltage design of SC-circuits is the realization of switches. Signal dependent switch resistances, clock feed-through and charge injection will cause distortion (see second chapter). Clock boosting [2] has been used to increase the overdrive voltage of the critical input switches, by means of standard charge pumps. The boosted clock is applied to dual-GOX devices. These devices introduce no extra costs because they are already required for digital interface requirements. The clock generator circuitry runs at 1.5 V supply voltage. The boosting is done right at the critical switches. These options permit to obtain well-behaving switches without using excessively large transistors, and keeping the power consumption and the area of the clock distribution circuitry quite small.

4.2.1 Operational-Transconductance-Amplifier

The Operational Transconductance Amplifier (OTA)'s are the major source of power consumption in the analog part of the modulator. Therefore, the right choice of the OTA architecture can resolve the *"power-accuracy-speed"* trade-off [71]. Employing deep-submicron technologies makes it easier to realize high speed OTA's. But, the OTA noise-level may be an issue. In this design, the noise-floor of the modulator is dominated by the contribution of the first integrator, and the noise of the OTA is not negligible. It is well known that single-stage OTAs have speed advantages with respect to multistage OTA (e.g. Miller OTA), although, they have the drawback of a reduced output signal-swing. Due to the feed-forward approach, which allows the reduction of the output swing capability of the OTAs, folded cascode OTAs could be employed, leading to superior power efficiency [71]. In fact, in Miller OTAs additional power is needed to obtain a good phase-margin. Extensive behavioral simulations resulted in specifications for the OTAs shown in Table 4.4.

Due to the fairly low intrinsic gain of deep-submicron MOS-transistors, nonminimum length transistors have been used to enhance the output conductance g_{ds}. Also the operation region of the transistors has been carefully selected to decrease g_{ds}. No special technique, like gain-boosting, has been used in order to reduce the power overhead. A schematic of the folded cascode OTA is illustrated in Fig. 4.4 employing switched-capacitor common-mode control. Care was taken for process variations, for supply tolerance (± 10 %) and for the temperature range (from -40 degrees up to $+125$ degrees Celsius) during the design phase.

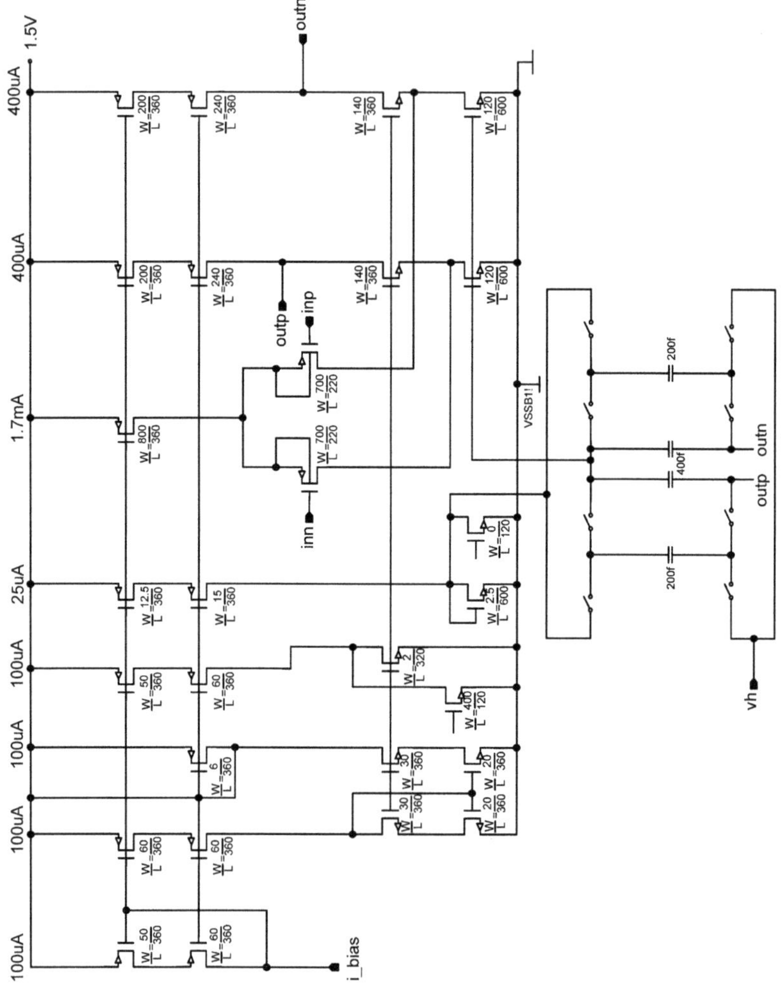

Fig. 4.4 Schematic of first OTA

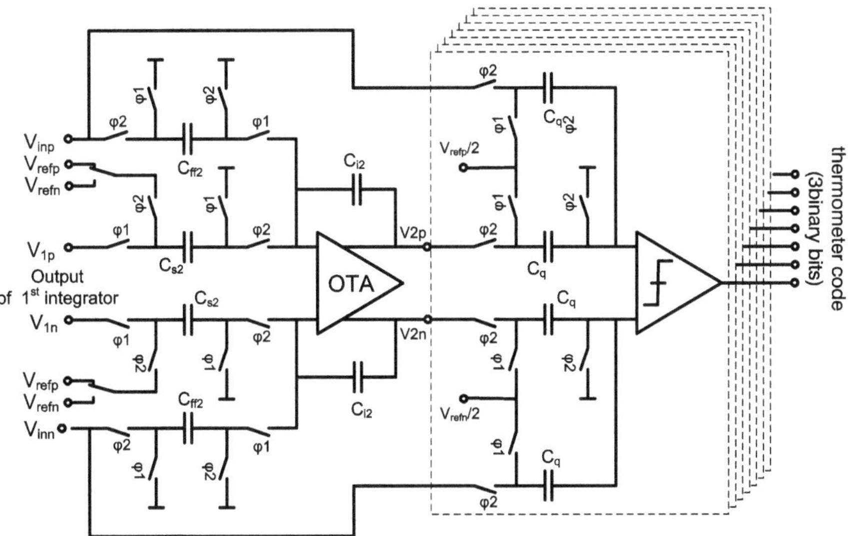

Fig. 4.5 Schematic of 2nd integrator and quantizer

4.2.2 *Second-Integrator and 3 Bit-Quantizer*

The feed-forward approach requires two additional adders as compared to the conventional delta-sigma modulator. The first adder is placed at the input of the second integrator, the second adder is needed right at the input of the quantizer. The feed-forward summing node at the input of the second integrator is realized by introducing an additional input to the second integrator, as shown in Fig. 4.5. ϕ_1 and ϕ_2 denote the odd and even phase of the non-overlapping clock, respectively. The capacitor C_{ff2} samples directly the signal at the input of the converter, whereas the capacitor C_{s2} samples the output of the second integrator. To relax the power consumption requirement of the second integrator the feed-forward input is sampled during the integration phase of the second integrator. The second feed-forward path at the input of the quantizer, is realized such that the signal at the input of the modulator and the output of the second integrator are sampled on two capacitors during the integration phase of the second integrator. Charge redistribution makes the voltage across this capacitors equal to $\frac{V_{o2}+V_{in}}{2}$. To compensate for this signal scaling by $\frac{1}{2}$ the quantizer has a virtual gain of 2. Thus, the input range of the quantizer is half of the reference voltage V_{ref}. This increases the precision required to the comparators.

V_{ref} was chosen to be 0.7 Vp, therefore the quantizer's LSB results in 50 mV, which permits to go for a straight forward circuit design. The proposed *passive adder* helps to keep the power drain low. Clock-boosting has been used to increase the overdrive voltage of the critical input switches, by means of standard charge pumps [2]. The boosting is done right at the switches to keep the power consumption of the clock distribution circuitry small.

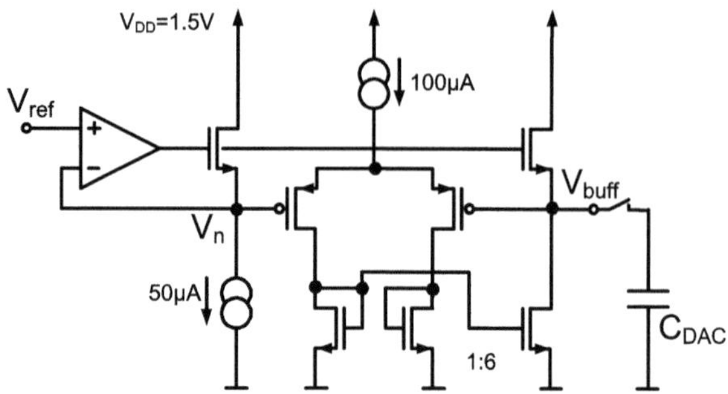

Fig. 4.6 Schematic of global open-loop voltage reference buffer

4.2.3 Reference Buffer in Open-Loop Configuration

The DAC's capacitors and input capacitors are shared, which permits to save area and power [75]. But, the charge delivered from the reference buffers is dependent on the input signal. This may lead to distortion, if sufficient settling is not ensured. Therefore, low-impedance open-loop buffers have been designed instead of a classical OpAmp's in non-inverting configuration. This leads to less power consumption because the global open-loop operation allows the use of the full-speed of the devices. Figure 4.6 depicts a schematic of the global open-loop reference buffer. The OpAmp and the NMOS transistor at its output makes a copy at node V_n of the reference voltage applied at the non-inverting input terminal. There are no high-speed requirements resulting in a very low-power drain of 50 μA for the OpAmp. The differential-pair balances the voltages at both input terminals providing a replica of the input voltage to the output (V_{buf}). The buffered voltage (V_{buf}) drives the capacitive load given by the D/A-converter. Closing the switch causes a large current spike at the output for charging the capacitor C_{DAC}. The current spike unbalances the circuitry and the differential pair steers more current through the output until the balanced state is reached again. Careful design showed that a tail-current of 100 μA is sufficient to provide a proper settling transient voltage across the DAC-capacitors.

4.3 Measurement Results

The discussed delta-sigma modulator was implemented in a 0.13 μm 4-metal standard digital CMOS-technology using a dual-GOX process option for the input switches. These devices do not introduce extra costs because they are available for digital chip-interface requirements. Clock-boosting is employed for achieving sufficient linearity of the input switches. Care was taken for occurring voltage spikes to comply with all reliability constraints. The chip photo of a test chip is depicted in Fig. 4.7. A significant part of the converter's silicon area needs to be spent for power routing. Maintaining a lifetime of 15 years and a temperature range from −40 degrees up to +120 degrees Celsius requires wider metal-lines to avoid electromigration issues.

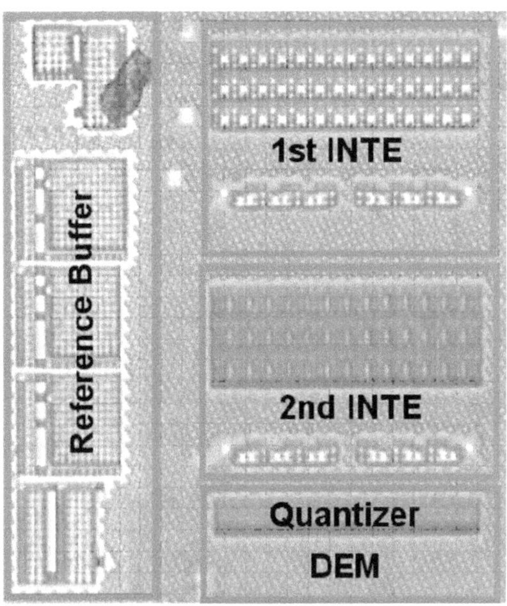

Fig. 4.7 Micro-graph of 0.13 μm delta-sigma modulator

The ADC core-area results in 0.14 mm^2, all required reference buffers consume additionally 0.15 mm^2 of chip-area. Clearly, the area is dominated by capacitors whereas transistors play a minor importance. As compared to the design in 0.18 μm [47] with similar performance, the silicon area decreases by 25 %. The linear capacitors were implemented as advanced sandwich devices reducing manufacturing complexity for MIM-capacitors at the expense of worse matching properties. Thus, for achieving sufficient linearity in both feed-back DACs, Dynamic Element Matching (DEM) techniques were implemented. An on-chip anti-aliasing filter drives the modulator, consuming 6 mW of power. The corner frequency was placed at 1.5 MHz for the second-order Butterworth-filter. All the bias-currents and reference voltages are derived from an on-chip programmable biasing-circuit including a bandgap reference circuit. Digital data are captured into an on-chip Random-Access Memory (RAM) at high sampling-rate. RAM data are read from the chip at a low frequency utilizing a logic analyzer. Post-processing is performed off-line using MATLAB. A 131072 points FFT plot is depicted in Fig. 4.8 (0 dBFS refers to 1.4 Vpp). Spectral averaging over 3 measurements is applied. The band of interest is dominated by white-noise. Starting at 1.5 MHz the second-order shaped-noise becomes dominant. The input tone at 100 kHz with an amplitude of −6 dBFS was applied, revealing a second- and third-harmonic component of 84 dBc and 82 dBc, respectively. Figure 4.9 shows the resulting signal-to-noise ratio (SNR) and SNR plus distortion (SNDR) versus the normalized input signal $\frac{V_{in}}{V_{ref}}$. Up to a normalized input signal of −6 dB the noise energy is dominant over the harmonics. Thus, the peak value for the SNR and SNDR is reached below a full-scale input signal.

Figure 4.10 illustrates a Discrete Multi-Tone (DMT) signal which is commonly used in ADSL applications. Each modulated tone carries a certain number of bits

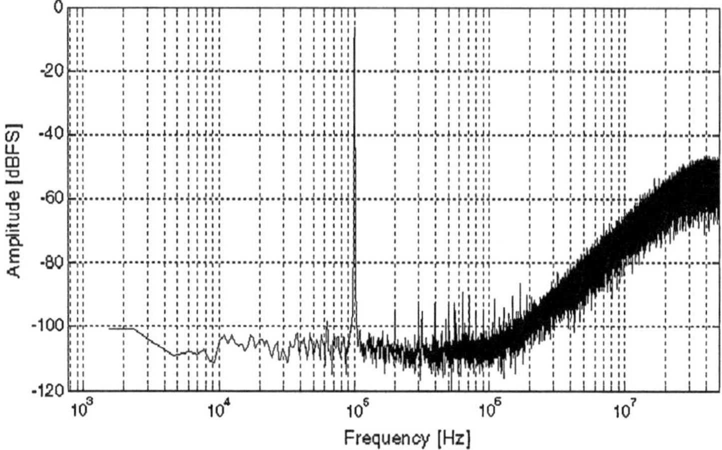

Fig. 4.8 Output data-stream PSD-plot (−6 dBFS)

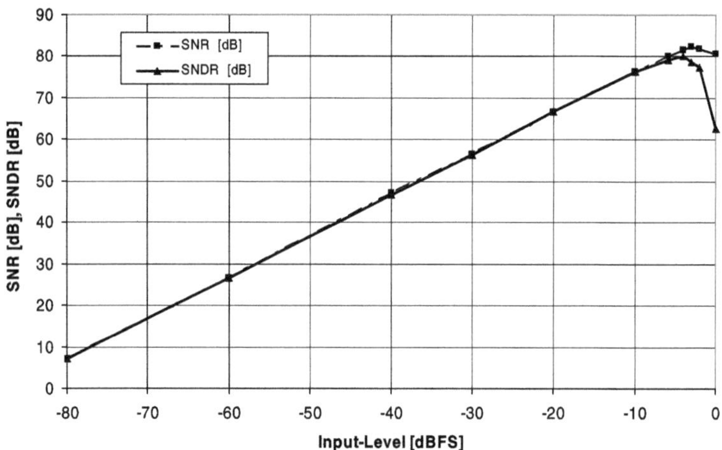

Fig. 4.9 Measured SNR and SNDR vs. input-level

for data transmission. The resulting peak-to-rms ratio (CREST-factor) of the DMT-signal varies due to the instant bit-allocation. The resulting CREST-factor can go up to 5.6 (15 dB) challenging the noise and linearity requirements. The Missing-Tone Power Ratio (MTPR) evaluates the overall linearity of an ADSL system. Hereby, one of the DMT-tones is omitted, as the fourth tone in Fig. 4.10 (see circle). The distance between the level of DMT-tones and the sum of all occurring intermodulation products at the omitted tone is bigger than 70 dB. Thus, the achieved linearity and noise contribution is sufficient for maintaining maximum data-rate. Table 4.5 summarizes some of the important measurement data of the designed feed for-

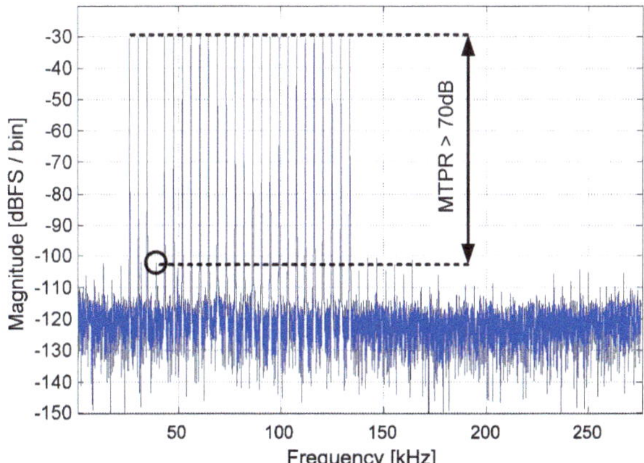

Fig. 4.10 Spectrum of DMT-signal with CREST = 5.6

Table 4.5 Measured performance summary of feed-forward A/D-converter

Operating mode	Data-transfer	Line-testing
Signal bandwidth	276 kHz	1.1 MHz
Sampling frequency	105 MHz	105 MHz
Oversampling ratio (OSR)	192	48
Reference voltage (diff.)	0.7 V	0.7 V
Dynamic range (DR)	88 dB	82 dB
Peak SNR	82 dB	78 dB
Peak SNDR	80 dB	76 dB
Power ADC core (1.5 V)	8 mW	8 mW
Power reference (1.5 V)	7 mW	7 mW
Pre-filter (1.5 V)	6 mW	6 mW
Decimation-filter (1.5 V)	3.1 mW	3.1 mW
Technology	0.13 µm 1.5 V	CMOS 4ML
FOM (DR) (ADC-core I ADC + Ref)	0.7 pJ/conv	1.3 pJ/conv
FOM (peak-SNR) (ADC-core I ADC + Ref)	1.4 pJ/conv	2.6 pJ/conv
FOM (peak-SNDR) (ADC-core I ADC + Ref)	1.7 pJ/conv	3.2 pJ/conv

ward A/D-converter. The measurements were done for both modes, the data-transfer (bandwidth = 276 kHz) and for line-testing features (bandwidth = 1.1 MHz).

4.4 Conclusion

The evolution of DSL technologies employs the cost efficient System on Chip (SoC) approach [42]. Digital design demands deep-submicron technologies in order to reduce power consumption and to enlarge the integrated functionality. For analog modules, migration to supply voltages below 3.3 V is more critical. Multi-chip packaging is not always a preferable approach since the total manufacturing costs are quite high. Taking these points into account it becomes necessary to build an Analog Front End (AFE) in a 0.13 µm technology, allowing the integration of the AFE together with digital blocks of the modem on one single silicon die.

In ADSL-CO applications the power consumption is one of the most important criteria [42]. This has a strong impact on architectural considerations. A high-resolution delta-sigma A/D-converter in 0.13 µm CMOS was presented [48]. The power drain and area consumption was reduced without deteriorating the targeted performance. Challenges in low-voltage circuit design have been addressed by an architectural approach. Usually Miller OTAs are used for such applications for their high output swing capability. Making use of a feed-forward topology single-stage OTAs could be used in the circuit, allowing an extremely power efficient implementation by trading-off between high SNR (large V_{ref}) and low distortion (small V_{ref}). The proposed solution introduces a *feed-forward*-approach of the input signal as depicted in Fig. 4.1. Both *feed-forward paths* c_1 and c_2 *bypass* the input signal so that inside the noise shaping loop the signal itself vanishes. The second feed-forward path at the input of the quantizer is realized by a *passive adder* employing charge redistribution techniques. The DAC's capacitors and input capacitors are shared, which permits to save area and power. This may lead to distortion, if sufficient settling is not ensured. Therefore, low impedance open-loop buffers (Fig. 4.6) have been designed instead of a classical OpAmp in non-inverting configuration. This leads to less power consumption because the open-loop operation allows the use of the full speed of the devices.

Compared to other implementations in larger feature-size CMOS processes [3], it is still possible to use standard circuits and still have advantages in terms of power dissipation and silicon area. Clocked at 105 MHz, the delta-sigma modulator and the reference voltage buffer achieve a dynamic-range of 88 dB and a peak-SNR of 82 dB consuming 8 mW and 7 mW, respectively, from a 1.5 V supply [48]. As compared to the design in 0.18 µm [47] with similar performance, the silicon area decreases by 25 % thanks to the improved architecture. Furthermore, the overall power drain could be reduced from 25 mW down to 15 mW resulting in a FOM (2.7) of 0.7 $\frac{pJ}{conv}$ and 1.3 $\frac{pJ}{conv}$ taking into account the power drain of the ADC and the ADC including all reference buffers, respectively.

Chapter 5
A Delta-Sigma Converter for WLAN Using a TEQ

Modern CMOS technologies employ the high-speed performances of nanoscale transistors. The possible supply voltage results in typically one Volts for a device consisting of a gate-oxide thickness of about 1.2 nm. One of the main challenges is the implementation of the multi-bit quantizer with an array of comparators resulting in a flash-converter. The performance of these converters is bounded by the tolerable limit for the power consumption of the flash-converter embedded in multi-bit architectures. This flash-converter is quite problematic to realize in low-voltage technologies, where the dynamic range of the comparators is limited and comparator offset degrades the linearity of the quantizer. Furthermore, a multi-bit D/A-converter is required in the feedback path whereas its impairments limit the overall performance.

Some alternatives have been explored to avoid use of direct amplitude quantization in A/D-converters, which are based on *time-encoding* rather than *amplitude-encoding* [9]. As already discussed in Chap. 2, an exchange of the amplitude-axis for the time-axis offer a possibility of overcoming resolution problems in A/D-conversion in low-voltage CMOS circuits. This exchange can be done by some form of *duty-cycle modulation* or *pulse-width modulation*. Conceptually, this means an amplitude to square-wave conversion by some square-wave modulation scheme, followed by a discretization of the time-axis. For its implementation a circuit configuration is used consisting of an *asynchronous* delta-sigma modulator, as depicted in Fig. 2.19. This approach is one of several possible solutions to avoid direct amplitude quantization. An innovative Time-Encoding Quantizer (TEQ) is introduced to overcome design issues in a 1.0 V supply-voltage 65 nm digital CMOS technology. The FOM as described in (2.7) results in 0.15 pJ/conversion-step. Furthermore, the total area of the $\Delta\Sigma$-modulator consumes 0.079 mm^2 of silicon area only.

An asynchronous $\Delta\Sigma$-modulator has already been used for data conversion. In [9] a sampler was placed outside the loop leading to an architecture which is rather different to this work. The generated binary output signal is discrete in amplitude but still continuous in time, since the duty-cycle represents an analog value. In order to get a uniformly sampled signal, the binary PWM-signal needs to be applied to a Time-to-Digital Converter (TDC). The basic task of a TDC is to quantize the

R. Gaggl, *Delta-Sigma A/D-Converters*, Springer Series in Advanced Microelectronics 39, DOI 10.1007/978-3-642-34543-2_5, © Springer-Verlag Berlin Heidelberg 2013

time interval between a start- (rising-edge) and a stop-signal (falling edge) and to provide a digital representation. The difference between a TDC and a simple counter is the high-resolution in the order of a few picoseconds [63]. In [10] the sampler was placed inside the loop so that it can be considered to be one of the first PWM combined with $\Delta\Sigma$-modulation. However, a complex poly-phase implementation of the sampler was proposed and the potential for this architecture to favorably replace a multi-bit quantizer inside a $\Delta\Sigma$-modulator was not yet considered. The approach for embedding a pulse-width modulator inside a Delta-Sigma modulator was later on introduced by [31] and [12] as well. An innovative research was reported in [58] and [59] showing that the PWM inside a $\Delta\Sigma$-modulator is comparable to a $\Delta\Sigma$-modulator operating at a limit-cycle and studying the properties of this limit-cycle.

In this chapter a novel architecture is presented employing a time-encoding quantizer implemented in a continuous-time multi-bit delta-sigma modulator, as basically introduced in [31]. The time-encoding quantizer replaces a flash A/D-converter and is based on a modulated oscillator producing a two-level signal. In this work the Time-Encoding Machine (TEM) proposed in [31] has been further simplified to avoid a multibit-DAC in the feedback loop. A TEM is reduced to a Time-Encoding Quantizer (TEQ) by using a pulse-width modulated binary-signal as feedback instead of a multi-bit signal. Basically, an analog signal is applied to a TEQ which generates a uniformly sampled binary signal representing the input amplitude by its discretized duty-cycle. Contrarily, a binary rectangular signal is applied to a TDC which provides a digital representation of the duty-cycle. This approach is adequate for nanoscale technologies and may permit a significant reduction of circuit complexity along with reduced power consumption of the flash-converter. Furthermore, the proposed solution can be scaled down for 45 nm, 32 nm and even smaller feature size technology nodes.

5.1 The Principle of a Time-Encoding Quantizer

Time-encoding is suitable technique for implementation in nanoscale technologies due to small gate-delays permitting sufficient time resolution. Derived from the classical Pulse-Width Modulation (PWM) modulators used in power electronics, research towards A/D-converters issued a number of efficient PWM publications [1, 9, 19] and [57]. All of them require a digital decoder clocked at very high frequencies to convert the PWM signal into a uniformly sampled multi-bit signal. The decoder employs the fact that the input signal spectrum is reproduced in the low frequency components of the PWM output signal, as shown in formula (2.49) and Fig. 2.22. The main technological barrier was a very high oversampling frequency required for the digital decoder making the approach of a PWM A/D-converter less attractive for real products.

The principle of a Time-Encoding Machine (TEM) was introduced in [1] referring to a continuous-time system encoding a band limited signal using a binary signal. In the paper of L. Hernandez [31] it is shown that a set of irregular samples [23]

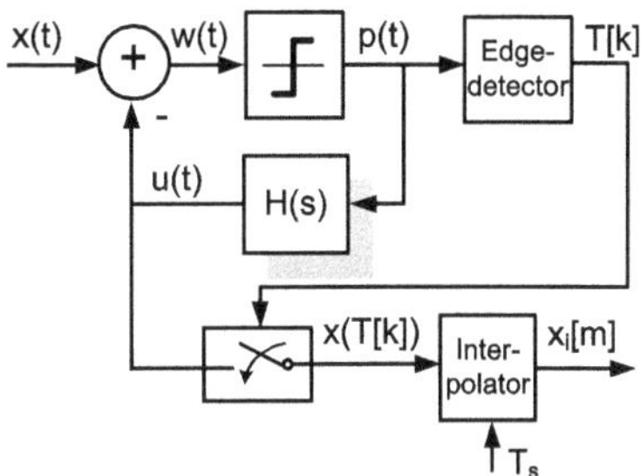

Fig. 5.1 A time-encoding machine (TEM)

of the input of a TEM can be extracted from its binary coded output. Combining a TEM and an irregular sampler with a loopfilter results in a possible implementation of a delta-sigma modulator. This modulator produces a *noise-shaped* and *uniformly sampled* multi-bit signal, similar to a conventional delta-sigma modulator. However, the modulator's TEM performs internally an irregular sampling and derive the multi-bit output from a binary signal. The principle of a TEM is explained in [31] based on a *self-oscillating* nonlinear system which encodes a continuous-time band limited input signal into a binary signal. Such a system allows extracting from the binary signal an irregularly sampled sequence which complies with the conditions to ideally recover the input signal [23]. The irregularly sampled sequence can be converted into a uniformly sampled signal by a simple interpolator at the expense of introducing an error in the interpolation.

A possible concept for a TEM is shown in Fig. 5.1 and is similar to a self-oscillating Pulse-Width Modulation (PWM) used in many class-D amplifiers. Constructing an autonomous system that generates signal $u(t)$ by itself using a filter $H(s)$ and a feedback loop around the threshold sampler. The filter $H(s)$ is designed such that with null input $x(t)$, signal $u(t)$ is a limit-cycle of period T_{osc} and $p(t)$ is a square-waveform of period T_{osc} and amplitude $\hat{P}$. It will be assumed that for any $x(t)$, sufficient zero crossings will occur to generate a proper irregular sampling sequence $T[k]$. This can be accomplished by designing $H(s)$ such that signal $u(t)$ has a peak-amplitude of $\hat{U}$ and choosing T_{osc} small enough to guarantee that the zero crossings of $u(t)$ will be faster than T_{sig} for any input signal, providing some degree of oversampling. An approximated linear analysis of the system shown in Fig. 5.1 was done in [31]. $u(t)$ may be assumed a sine-wave of period T_{osc} that results from filtering $p(t)$ with the loopfilter $H(s)$, which usually has a lowpass characteristic and will attenuate the harmonic components of $p(t)$. The amplitude $\hat{U}_{sin}$ of this sine-wave may be estimated as follows:

$$\hat{U}_{sin} \cong \hat{P} \cdot \left| H(\jmath\omega_{osc}) \right|, \quad \omega_{osc} = \frac{2\pi}{T_{osc}}. \tag{5.1}$$

Assuming $x(t)$ is a tone of frequency ω_{sig} an approximated bound for the maximum input tone amplitude A_{max} can be derived as follows:

$$A_{max} < \hat{U}_{sin} \cdot \left| 1 + H(\jmath\omega_{sig}) \right| \cong \hat{P} \cdot \left| H(\jmath\omega_{osc}) \right| \cdot \left| 1 + H(\jmath\omega_{sig}) \right|. \tag{5.2}$$

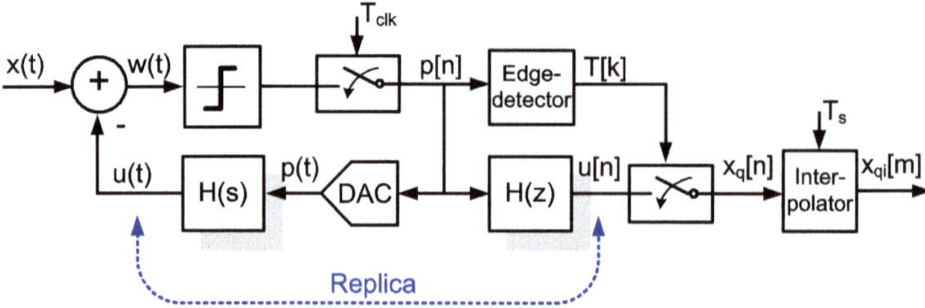

Fig. 5.2 A time-encoding quantizer (TEQ)

Figure 5.1 shows an interpolator producing a sequence $x_i[m]$ by computing approximated values of $x[m]$ from the irregularly sampled sequence $x(T[k])$.[1] Although a perfect interpolation may be possible [9], a restriction to simple interpolation algorithms is preferred at the expense of introducing an interpolation error $e[m]$. One of the simplest interpolation methods is chosen which is the causal nearest neighbor interpolation. Although the error $e[m]$ is not purely random for a sinusoidal input, it can be assumed that it behaves as a random process with uniform distribution [31]. Hence, its variance can be expressed as:

$$\sigma_e^2 = \frac{1}{12}\left(2 \cdot \frac{T_{osc}}{C} \cdot A_{max} \cdot \omega_{sig}\right)^2, \qquad (5.3)$$

whereas $C = 1$ if $u(t)$ is sampled only with one edge polarity and $C = 2$ if $u(t)$ is sampled with the rising and falling edges.

The outcome of the TEM in Fig. 5.1 is of little use for implementation in an A/D-converter, since the generated output signal is not suitable for further digital signal-processing.[2] To obtain a uniformly sampled sequence of quantized samples the proposed TEM will be adapted properly resulting in a so called Time-Encoding Quantizer (TEQ), as shown in Fig. 5.2. The modulator in Fig. 5.1 was modified by introducing a uniform sampler within the loop with sampling period T_{clk}. Thus, a uniformly sampled sequence of quantized values is generated, which can be handled by digital circuits.

The binary sequence $p[n]$ is applied to a digital filter $H(z)$ to generate $u[n]$, which is a sampled replica of signal $u(t)$. Reproducing the effect of filter $H(s)$ with the digital filter $H(z)$ whose impulse response is defined by imposing the impulse invariance principle to filter $H(s)$ at a sampling rate T_{clk}:

$$h_a(t) = L^{-1}\{H(s)\},$$
$$H(z) = Z\{h_d[n]\}, \quad h_d[n] = h_a(n \cdot T_{clk}). \qquad (5.4)$$

---

[1] Note: $x(T[k]) = u(T[k])$ at occurring edges.

[2] Note: $u(T[k])$ is an irregularly sampled sequence requiring an interpolator which is a non-conventional digital building block.

Including the sampler within the loop (Fig. 5.2), the values of $u(t)$ in the sampling points $(u(nT_{clk}))$ are for sure equal to $u[n]$, the output signal of filter $H(z)$. Note, including the sampler in the loop may change the conditions of oscillation of the system which now turns into a *discrete-time dynamical system*. The generated discrete-time limit-cycle should be a sampled version of the continuous-time limit-cycle of the original system depicted in Fig. 5.1. Defining the ratio between the sampling frequency $f_{clk} = 1/T_{clk}$ and the nominal limit cycle frequency $f_{osc} = 1/T_{osc} = \omega_{osc}/(2\pi)$ by the so called Ratio-OSR (ROSR):

$$ ROSR = \frac{f_{clk}}{f_{osc}} = \frac{T_{osc}}{T_{clk}}, \quad ROSR > 2. \tag{5.5}$$

The discrete limit-cycle and the continuous limit-cycle can be made almost equal if ROSR is sufficiently large and $H(s)$ consists of an asymptotically lowpass characteristic and monotonic phase [31]. As shown in Fig. 5.2, the sequence $x_q[n]$ is generated by extracting a subset of the sequence $u[n]$ controlled by the edge detector. Thus, the irregularly sampled sequence $x(T[k])$ of Fig. 5.1 gets resampled accordingly establishing some relationship between both systems. However, time quantization will add an additional error in the system since the zero crossings of signal $p[n]$ in Fig. 5.2 do not correspond exactly with the sampling points of signal $p(t)$ produced in the system of Fig. 5.1. A detailed analytical analysis reveals following estimated variance σ_q for the occurring time quantization error:

$$ \sigma_q^2 \approx \frac{4}{3} \cdot \left(\frac{C \cdot \hat{P}}{ROSR} \right)^2. \tag{5.6}$$

In case of sampling with one edge or with both edges (rising and falling), $C = 1$ or $C = 2$, respectively. $\hat{P}$ denotes the amplitude of the pulse $p(t)$. Note, that sampling in both edges produces a larger time quantization variance. The system in Fig. 5.2 can now be used as an analog-to-digital converter. Establishing an approximated relation between the design parameters such as ROSR and the limit-cycle period T_{osc} leads to the overall error resulting from the difference between $x_{qi}[m]$ and $x[m]$, the outcome of uniformly sampling $x(t)$ with a sampling period T_s. Assuming a sinusoidal input signal of frequency ω_{sig}, the total variance σ_t of the error can be approximated [31]:

$$ \sigma_t^2 \approx \sigma_q^2 + \sigma_e^2 = \frac{4}{3} \cdot \left(\frac{C \cdot \hat{P}}{ROSR} \right)^2 + \frac{4}{3} \cdot \left(\frac{\pi}{C} \cdot A_{max} \cdot \frac{\omega_{sig}}{\omega_{osc}} \right)^2. \tag{5.7}$$

The approach of the proposed time-encoding quantizer can be used for an analog-to-digital conversion. It has been shown that the time-encoding quantizer introduces two error components, one due to the time quantization and another due to the interpolation process. Keeping the implementation simple, a nearest neighbor interpolator was taken into account.

5.2 System-Design Considerations

Continuous-time delta-sigma modulators are often employed as A/D-converters. These modulators are an attractive approach to implement high-speed converters in VLSI systems due to the benefits of a feedback configuration they have low sensitivity to circuit imperfections compared to other solutions such as pipeline- or flash-converters. In Sects. 2.3.6 and 2.3.5 a short overview is given about the analysis, modeling and design of high-speed continuous-time delta-sigma modulators. The resolution and the stability of these modulators are limited by two main factors, excess-loop delay and sampling uncertainty. Both factors, among others, have been carefully analyzed and modeled in [24].

5.2.1 Integration of a TEQ into a Delta-Sigma Modulator

Replacing a conventional coarse quantizer such as a flash-converter by a Time-Encoding Quantizer (TEQ) can be done in a straightforward manner. A TEQ results in a very simple hardware implementation as compared to a similar flash-converter. However, attention needs to be paid to the basic differences between a flash-converter and a TEQ. Firstly, the Dynamic Range (DR) of a flash-converter is determined by a known reference voltage V_{ref}, whereas the TEQ shows a DR that depends on the spectral components ω_{sig} of the input signal (5.2). Secondly, a flash-converter gets triggered periodically with a sampling clock. Contrarily, the sampling instant of a TEQ is triggered by a signal event which does not correspond exactly with uniform sampling intervals required by the feedback D/A-converter. Thirdly, the flash-converter can be seen as a quantizer generating a signal independent noise fulfilling the white-noise model (2.4). In contrast, a TEQ exhibits two conversion errors, a signal dependent interpolation error (5.3), and a signal independent resampling error (5.6).

Taking these three main differences into account, leads to following design process, as proposed in [31].

- *Gain Scaling*
  The maximum tolerable input amplitude to the TEQ depends on both the transfer function $H(s)$ (see Fig. 5.2) and on the ratio between the signal bandwidth ω_{sig} and the limit-cycle frequency ω_{osc} (5.2). Exploiting the full DR, a gain coefficient k_g has been introduced at the input of the TEQ, which can be approximated as:

$$k_g \approx \hat{P} \cdot \left| H(J\omega_{osc}) \right| \cdot \left| 1 + H(J\omega_{sig}) \right|. \tag{5.8}$$

  Note, maintaining the desired state variable at the output of the feedback D/A-converter, a scaling of $1/k_g$ is required additionally.

- *Relation Between ω_{osc} and ω_s*

  Defining COSR as the ratio between the limit-cycle frequency ω_{osc} and the sampling-frequency ω_s of the CT delta-sigma modulator yields:

  $$COSR = \frac{\omega_{osc}}{\omega_s} = \frac{f_{osc}}{f_s}. \tag{5.9}$$

  Keeping the speed requirements low for the active building blocks, ω_{osc} should be as small as possible. For achieving a sufficiently dense sampling sequence $u[n]$ in Fig. 5.1, the minimum limit-cycle frequency results in $COSR \geq 1/C$. The irregular sampling process introduces a random time varying delay between the actual sampling instant and the update of the feedback D/A-converter. Ensuring a stable modulator-loop a maximum tolerable excess-loop delay needs to be defined. Usually, this is given by a fraction E of the sampling-period $T_s = (2\pi)/\omega_s$. Thus, the lower bound for $COSR$ is given by

  $$COSR \geq \frac{1}{C \cdot E}. \tag{5.10}$$

- *Relation Between Quantizer-Resolution and ROSR*

  The derived noise model (5.7) is valid if a sinusoidal input signal is applied. Imposing the same SNR for a uniform sampling quantizer and an equivalent TEQ yields a value for ROSR. Considering a TEQ connected within a delta-sigma modulator loop reveals that the interpolation error is smaller than predicted in (5.3), since this model is valid only for sinusoidal input signals. Resulting in a noise contribution of a uniform quantizer with N-bits of resolution, the minimum ROSR for the equivalent TEQ can be expressed as:

  $$ROSR_{min} \geq 2 \cdot C \cdot \hat{P} \cdot \frac{2^N}{k_g}. \tag{5.11}$$

The choice of the gain-scaling coefficient k_g, the ratio between the limit-cycle frequency ω_{osc} and the sampling frequency ω_s ($COSR$) and the $ROSR_{min}$ completes the design procedure of converting a conventional delta-sigma modulator into a delta-sigma modulator with a Time-Encoding Quantizer (TEQ).

5.2.2 Design-Specifications and System-Level Design

For Discrete Time (DT) modulators, implemented in a switched-capacitor technique, the analog bandwidth is typically in the range of a few MHz. However, attempts have been reported to enlarge the bandwidth of ADCs by going from Switched-Capacitor (SC) circuits to Continuous Time (CT) circuits, achieving a high bandwidth in the range of 30 MHz. Although it might be feasible to go for a SC-solution in nanoscale technologies, for the targeted WLAN application a Continuous

Table 5.1 System specification for WLAN A/D-converter

ENOB (nominal)	10 bits
ENOB (worst case)	9.5 bits
Analog bandwidth	70 kHz to 18.5 MHz
Technology/supply voltage	65 nm CMOS/1.0 V

Time (CT) modulator was considered adequate due to its beneficial low-power consumption. This is because a CT A/D-converter requires a much lower ratio between GBW and sampling-frequency as compared to equivalent SC counterparts [24]. Besides the advantage of CT circuitry with respect to higher achievable bandwidth or lowered power dissipation, there are some difficulties to overcome as discussed in Chap. 2. First, the development of a suitable architecture is more challenging than in an SC design approach since the CT components show a higher variation with process, temperature, and supply-voltage spread. Second, the CT feedback circuitry leads to a less stable loop due to the Excess Loop Delay (ELD). Finally, the internal feedback D/A-converter is much more susceptible to clock-jitter than its SC counterpart.

Table 5.1 shows the required system specification for a WLAN A/D-converter supporting IEEE 802.11n (Table 1.2). For achieving the targeted resolution of 10 ENOBs a well balanced noise budget has been considered for this purpose. The thermal noise will limit the overall performance at lower frequencies while shaped quantization noise will dominate above 20 MHz. In the analog bandwidth, the jitter contribution is assumed to be lower than -72 dB while the shaped quantization noise results in -69 dB. Thermal noise contributions of resistors and OpAmps may not exceed -65 dB. The final considered SNR results in 63 dB, which is 3 dB better than specified providing a half-bit of safety margin in the implementation.[3]

At system level, the resolution achieved by such a system is a function of the oversampling ratio (OSR), noise shaping order (L) and quantizer resolution (N), as shown in formula (2.44). Using Schreier's toolbox [53], several discrete-time NTFs have been evaluated finding a good trade-off among OSR, loopfilter order or noise-shaping order, quantizer resolution and robustness. To support the design process, a software tool was developed which supports a semiautomatic design and interactive optimization of the CT modulator as introduced in [60] and applied in [61]. The preferred NTF was constrained for an infinity-norm[4] of 1.5 providing good ro-

---

[3]The WLAN A/D-converter has been developed in a collaboration project between Infineon Technologies Design Center of Villach, Austria and Carlos III University, Madrid, Spain. The design of the modulator has been split in two main working packages. The system-level design was done by Microelectronics Group at Carlos III University, and circuit-level design (circuit specification, design of all stages at transistor level and layout) was done at Infineon Technologies. This A/D-converter was presented at ESSCIRC 2010 conference [8] and published in IEEE Journal of Solid-State Circuits [7].

[4]In LTI-systems, the infinity-norm is defined as the peak-gain of a frequency response:

$$\|H(s)\|_{\infty} = \max |H(\iota\omega)|.$$

Table 5.2 Chosen system parameters for WLAN A/D-converter

Infinity norm of NTF	1.5
Order of noise-shaping (L)	3
Quantizer resolution (N)	4 bits
OSR/sampling-frequency	16/640 MHz
ELD	0.5 (640 MHz)
Analog bandwidth	20 MHz

bustness and keeping the resolution for the DT modulator quite similar to the CT modulator after applying the time-invariant transform [60]. The achievable SNR for a very aggressive infinity norm of 3 would have been better by more than one bit, but maintaining high robustness was considered to be more important. The tolerated ELD was assumed to be 25 % of the sample period. The chosen architecture for the CT delta-sigma modulator is summarized in Table 5.2.

Until know a conventional CT delta-sigma design was approached. The modulator will be modified to incorporate a time-encoding quantizer (TEQ) by replacing a coarse flash-converter by a Time-Encoding Quantizer (TEQ). The resolution of the TEQ will be equivalent to the original 4-bit quantizer, thus the complexity of the analog circuitry may be significantly reduced. The simplified block diagram of this modulator design is depicted in Fig. 5.3. The TEQ has been further simplified as compared to Fig. 5.2. Instead of tapping the output signal $u[n]$ of the digital filter $H(z)$, the comparator output signal $p[n]$ is directly fed back into the delta-sigma loop. This signal contains the input signal and the generated limit-cycle (see Fig. 2.22), which is far above the signal bandwidth. This setup is similar to a PWM-modulator as described in Chap. 2 and will also encode the input signal producing a satisfactory SNR.

In contrast to an asynchronous PWM, the proposed solution inserts a sampler inside the pulse-width modulator to get a uniformly sampled signal. The input signal is applied to a CT loopfilter $H_{ds}(s)$ along with a feedback signal generated by the second D/A-converter DAC_2. The output signal of the loopfilter is quantized by the TEQ. This TEQ produces a fast sampled signal by means of pulse-width modulation using a latched comparator and an integrator. The *binary* output signal of the TEQ is directly fed to both D/A-converters. As compared to [31], another option is proposed, which replaces the digital filter $H(z)$ and the multi-bit DAC by a single-bit DAC followed by an analog filter. The analog reconstruction filter converts the binary pulse-width modulated signal into a equivalent multi-bit signal. Furthermore, the analog reconstruction filter serves as a tracking filter introducing gain in the feedback loop. This feedback gain reduces the amplitude at the input of the TEQ, which keeps the harmonics contributed by the TEQ sufficiently small. The

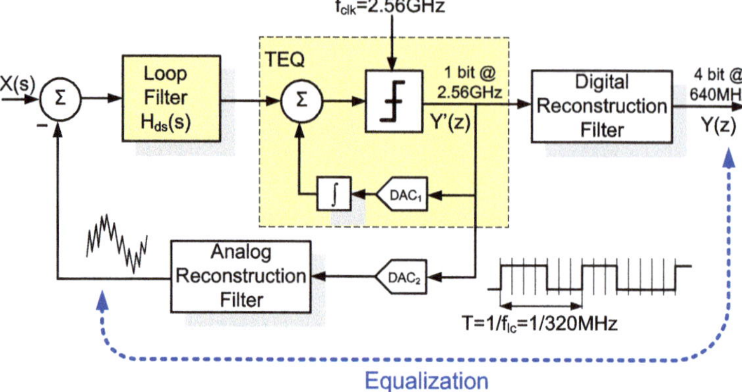

Fig. 5.3 Integration of a TEQ into a delta-sigma modulator

digital reconstruction filter provides a replica of the analog multi-bit signal. Note, this digital filter is outside of the noise-shaping loop.[5]

In the outer-loop (DAC_2) the lowpass filter reduces the quantization noise of the single-bit data-stream along with the spectral components associated with a PWM modulated signal, as given by (2.49). The filtered signal can be considered as the multi-bit signal of the equivalent 4-bit quantizer. As a penalty, the corner frequency of the lowpass filter needs to be placed slightly above the analog bandwidth, thus, the capacitor value gets relatively large.

The basic idea of the new implementation of Fig. 5.3 is that in Fig. 5.2 the signal $x_{qi}[m]$ is replaced by $p[n]$ for feeding the D/A-converter used in the delta-sigma loop. Thus, signal $u(t)$ (Fig. 5.2) is used in the feedback-loop of the modulator. That means, the time quantization error σ_q is spectrally shaped whereas the interpolation error σ_e is not shaped since it is outside of the delta-sigma loop. In practice, a relatively simple interpolation algorithm such as a polynomic interpolator may produce an interpolation error that has a negligible power compared with σ_q.

5.2.3 *Architecture for A/D-Converter*

The loopfilter has been optimized until it showed sufficient conditions for robustness against process variations, excess-loop delay and clock jitter, as shown in [60]. The resulting modulator out of the optimization process was simulated in DT to assess the final performance in terms of stability and SNR. The optimization was carried out by a MATLAB tool as proposed in [61]. The loopfilter poles were spread over

---

[5]As depicted in Fig. 5.2, the modulated single-bit signal is applied to a digital filter $H(z)$ which is converted into the desired multi-bit signal. In [31], the multi-bit data are fed back by a *multi-bit* DAC in the outer-loop calling for linearity and noise being better than the overall performance.

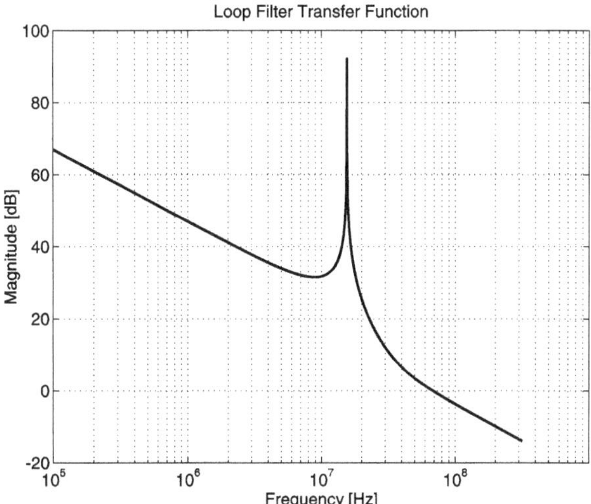

Fig. 5.4 Loopfilter transfer function for WLAN A/D-converter

the analog bandwidth according to Schreier's toolbox [53] and two additional loop-filter zeros were placed to stabilize the system. Circuit impairments were included with this methodology from the beginning of the design process in the CT domain. On the other hand, stability was checked by DT transient simulations to reduce simulation time. The discrete-time NTF was derived by computing it with the impulse invariant transform (2.42) of $H_{ds} \cdot H_{DAC}(s)$, whereas a non-return-to-zero DAC was assumed [24]. The loopfilter has been optimized to tolerate uncompensated ELD between 50 % and 75 % of sample period T_s. It has one pole at DC, two complex conjugated poles at 15.5 MHz (infinity Q) and two complex conjugated zeros at 29 MHz ($Q = 0.7$).

The transfer function of the analog loopfilter is depicted in Fig. 5.4, whereby the zero-dB crossing occurs at 68.13 MHz revealing a slope close to -20 dB/decade.

There are several topologies that can accommodate the loopfilter transfer function as shown in Fig. 5.4. Generally, a loopfilter can be implemented in a feed-forward or feedback configuration [54]. The simplified block diagram of the chosen architecture is depicted in Fig. 5.5. A multiple feedforward configuration has been chosen to keep the silicon area small, since each branch of a multiple feedback configuration would require one D/A-converter. A capacitive adder is used for the summing node for all feedforward branches to avoid a power penalty due to one additional active stage. The targeted loopfilter provides a resonator consisting of the first and the second integrator and the local feedback coefficient g. A conventional third integrator is cascaded to the resonator acting auxiliary as capacitive adder for both feedforward branches. Both feedforward branches comprise the capacitive adder ($a_1 \cdot s$) and ($a_2 \cdot s$) together with the third integrator (c_3/s), resulting in a virtual adder placed at the output of the whole loopfilter. The system coefficients depicted in Fig. 5.5 have been scaled properly keeping the maximum voltage swing in the range of 25 % of the supply voltage ($V_{DD} = 1.0$ V). Performing this scaling,

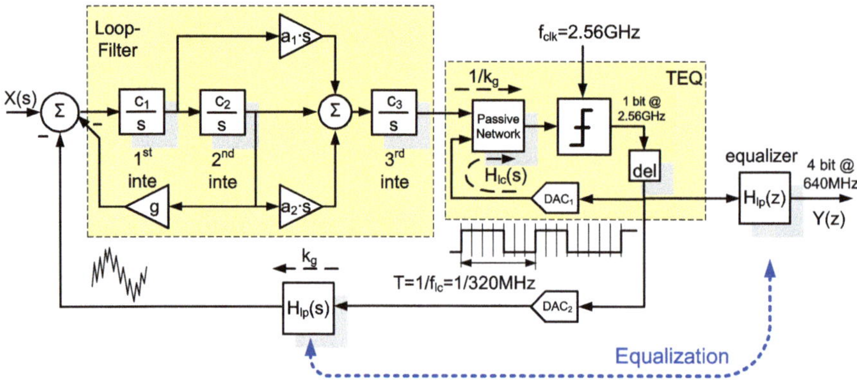

Fig. 5.5 Architecture of A/D-converter for WLAN

transient simulations on system level were carried out with a tone close to full-scale located at 20 MHz.

Both transfer functions of the TEQ are defined to be the same, in the inner-loop $H_{lc}(s)$ and in the outer-loop $H_{lp}(s)$. The transfer function of the lowpass filter is defined by one real pole ω_{lp} and a gain of g_{lp}:

$$H_{lp}(s) = H_{lc}(s) = g_{lp} \cdot \frac{\omega_{lp}}{s + \omega_{lp}} \tag{5.12}$$

whereas the gain g_{lp} of the lowpass is defined as:

$$g_{lp} = \frac{K_c}{k_g \cdot \omega_{lp}} \tag{5.13}$$

The parameters for the TEQ will be calculated for working on the rising clock edge only, e.g. $C = 1$. By means of (5.10), $COSR \geq 2$, e.g. the limit-cycle frequency results in $f_{osc} = 320$ MHz, which is half of the delta-sigma sampling-frequency. Choosing a limit-cycle oversampling ratio of $ROSR = 8$ fulfills well the condition (5.11), which leads to a clock frequency of $f_{clk} = 2.56$ GHz for the comparator. All those parameters were found in this design to match the target SNR by finding the 4-bit equivalent time-encoding machine for the given OSR and noise shaping order. The voltage based D/A-converter and the provided reference voltage V_{ref} define the full-scale voltage V_{fs} of the A/D-converter resulting in 160 mV.

5.3 Circuit Design

5.3.1 Loopfilter

Noise and linearity of the entire modulator are mainly determined by the first integrator. Integrator requirements decrease from the first stage to the last one because,

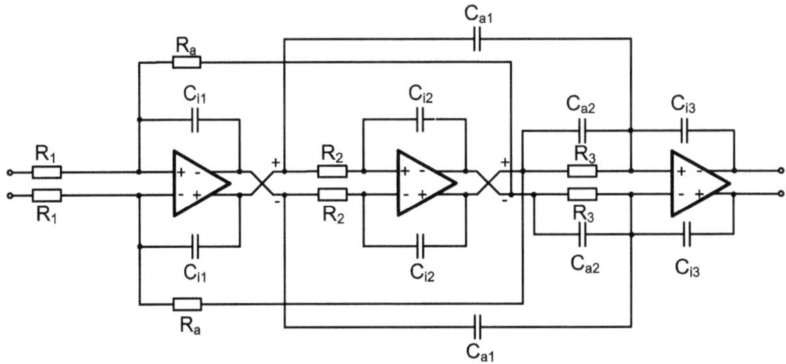

Fig. 5.6 Schematic of loopfilter for WLAN A/D-converter

when referred to the modulator input, any nonideality in one integrator is shaped
and attenuated by the transfer function of the preceding stages. There are two main
technologies to implement a CT integrator in CMOS technologies, the active-RC
and the $g_m C$ ones. In this work, an active-RC implementation will be considered,
since RC-filters are preferred in A/D-converters with multi-bit quantization and a
low OSR to achieve a high performance in a limited signal bandwidth. In contrast,
when single-bit conversion in a large bandwidth with a rather moderate accuracy is
demanded, $g_m C$-filters prove to be the power-efficient choice [50].[6]

Figure 5.6 shows the schematic of the CT loopfilter for the delta-sigma modulator. A multiple feedforward configuration has been chosen to reduce silicon area,
and the capacitive adder has been included to save power.

The capacitive adder comprises both feedforward branches C_{a1} and C_{a2} along
with the third integrator stage (C_{i3}). The capacitances C_{a1} and C_{a1} are given by
the equations $C_{a1} = a_1 \cdot C_{i3}$ and $C_{a2} = a_2 \cdot C_{i3}$, respectively. The value of R_1 is
constrained by the thermal noise contribution since its noise will be amplified by the
$|STF(j\omega)|^2$ imposing a limit to the achievable SNR. The thermal noise contribution
of R_2 and R_3 will be shaped by first and second order, respectively. Thus, both
resistors do not strongly influence the SNR, therefore, their values can be chosen
such that C_{i2} and C_{i3} result in a small capacitor value for the second and third
integrator. The resonator comprises the first two integrator stages together with the
feedback resistor R_a, whereas its value is defined by the ratio R_1/g. The succeeding
passive filter (see Fig. 5.5) will introduce a damping of $1/2$ in the signal path, thus,
c_{i3} will be halved to compensate this damping by an additional gain of 2 in the third
integrator.

---

[6]In case of a 65 nm CMOS technology, a signal bandwidth of 20 MHz is not yet considered as
large, whereas a SNDR of better than 10 ENOBs is already rated as high performance.

5.3.2 Integrator-Stage and Operational-Amplifier

For high frequency filters the g_m-stage in an $g_m C$-integrator can be reduced to its simplest possible configuration leading to a differential pair, to which some linearization techniques can be applied [71]. Due to its simplicity, these circuits are able to operate to very high frequencies. However, maintaining full-swing at the output to optimize the dynamic-range results in increased distortion components, due to the nonlinearities in the g_m-stage and basically to the nature of an open-loop structure. As a result this structure is only conditionally suitable for low-voltage design. Generally speaking, for linearity requirements up to about $THD = -50 \ldots -60$ dB, $g_m C$-filters are favorable, if low power is a major interest [35]. The targeted SNDR is at least 60 dB, that means another circuit such as an active RC-integrator is the preferred approach to take some design margin into account.

High resolution delta-sigma converters require a highly linear integrator such as an active RC-integrator (see Fig. 2.16) comprising an OpAmp in local-feedback configuration. This approach is preferable to open-loop structures, such as $g_m C$-integrators, in case of low supply voltage and tight distortion specifications. Furthermore, under worst-case conditions, the embedded supply voltage on block level can be 0.9 V only. The nominal supply voltage for the 65 nm core-device is specified at 1.0 V ± 5 %, the voltage drop across the on-chip supply rails may not exceed 50 mV. These circumstances lead to a multistage amplifier such as a two-stage Miller-OpAmp as the preferred approach. A two-stage cascade amplifier simply consists of two consecutive single-stage amplifiers showing the same DC-gain as compared to a cascode amplifier [71]. The two-stage amplifier comprises two high-ohmic nodes resulting in a two-pole system. To overcome stability problems, a Miller capacitor is added across the second stage to maintain sufficient phase-margin. This kind of frequency-compensation leads obviously to an increased power consumption. Choosing a single transistor for the output-stage permits an output swing of about $V_{DD} - 0.3$ V. In case of a cascode amplifier the possible output swing would be $V_{DD} - 0.6$ V resulting in an unsuitably small swing. The increased swing of a Miller-OpAmp allows to go for an increased thermal-noise contribution whereas the SNR can be kept the same. This fact helps to choose bigger resistance-values and therewith lowering the overall power drain. On top of that, the chip area can be reduced as well, since for the same RC-time-constant the capacitors of the integrators become smaller. However, an OpAmp in closed-loop configuration is a low-frequency building block as compared to a g_m-stage. Nevertheless, nanoscale technologies offer high-speed devices and a Miller-OpAmp can still provide sufficient gain at higher frequencies.

In a multiple feedforward configuration the bandwidth requirement of the OpAmps can be optimized as follows. The first integrator has to be the fastest and the last integrator can be very slow as compared to the sampling-frequency f_s. However, the capacitive adder imposes a certain speed limit on the last integrator, and makes it difficult to reduce the bandwidth of the last OpAmp to save power. The speed limitation of the last integrator depends on the GBW of the OpAmp, the unity

Table 5.3 Specification of Miller-OpAmps

	GBW [MHz]	DC-Gain [dB]	C_c [fF]	I(VDD) [mA]
$OpAmp_1$, $OpAmp_2$	700	45	520	1.4
$OpAmp_3$	855	45	450	1.5

gain frequency of the integrator and the whole capacitive load including feedforward branches and parasitics at virtual ground node (2.41). The specifications for the OpAmps have been verified by both mathematical modeling and transient simulations. It was found convenient to use the same specifications every OpAmp except for the last, due to the speed limit that the capacitive adder imposes to the system. The feedback path of every OpAmp includes a resistor R_x in series with the integrating capacitor to speed up their transient response, as shown in Fig. 5.10. The series-resistor R_x helps partially to overcome the effects due to limited GBW, as discussed in Chap. 2. The finally chosen values for the first, second and third integrator result in 1100 Ω, 600 Ω and 850 Ω, respectively. Table 5.3 gives an overview about the specifications of all three OpAmps. The given numbers correspond to the nominal case, whereas worst-case settings were taken into consideration during the design process as well. It is worth mentioning that all OpAmps are fully differential circuits with common-mode feedback loops providing a common-mode voltage of 0.5 V.

5.3.3 Time-Encoding Quantizer

The Time-Encoding Quantizer (TEQ) is based on the approach as shown in Figs. 5.3 and 5.5. Basically, the proposed TEQ comprises a single-bit comparator, a tunable delay-stage combined with an analog latch, a single-bit D/A-converter and a tunable lowpass filter. These building blocks are put in a single-loop configuration as shown in Fig. 5.7. The main purpose is to generate the self-oscillating frequency $f_{osc} = 320$ MHz for providing the limit-cycle signal. The generated limit-cycle will be modulated by the output signal of the loopfilter which is the input signal for the TEQ, denoted by V_i. The sum of the limit-cycle signal and the loopfilter output signal is done by a passive adder to save power. The comparator is clocked at a high clock-rate of $f_{clk} = 2.56$ GHz providing sufficient oversampling of the generated pulse-width modulated signal within the self-oscillating loop. The single-bit output signal V_{co} is latched by the non-inverted clock-signal to ensure signal integrity in the presence of inconstant comparator decision time. Note, the smaller the input signal to the comparator, the longer becomes the comparator decision time. A tunable delay has been inserted to control the delay within the loop for placing the self-oscillating frequency at the desired value of $f_{osc} = 320$ MHz. The latched signal is converted by a voltage-based D/A-converter back into the analog domain. The D/A-converter is directly biased from both supply-rails avoiding an additional

Fig. 5.7 Circuit for a TEQ

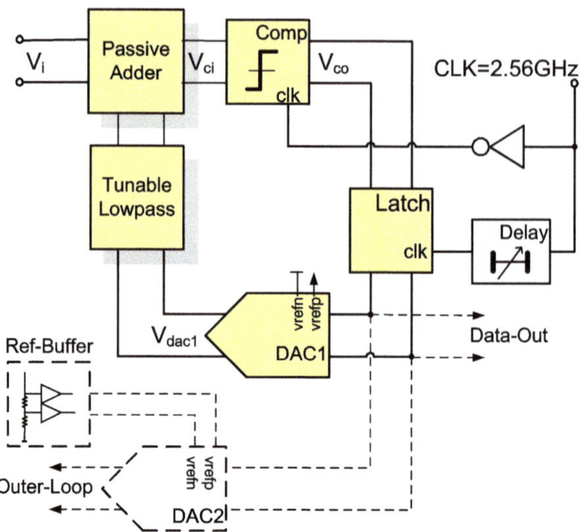

voltage reference buffer. Any injected noise and voltage deviation from the supply
rails does not degrade the overall performance, since any error will be third-order
shaped by the delta-sigma loop.

The analog signal V_{dac1} is applied to a tunable RC-lowpass filter containing a
real pole slightly above 20 MHz. A second pole at a higher frequency is implemented for tuning the phase-shift accordingly to set the self-oscillating frequency at
320 MHz. Thus, the spreading of the RC-time-constants can be compensated by the
programmable delay and the tunable phase-shift. Furthermore, the transfer function
of the passive adder from the input (V_i) to the comparator (V_{ci}) provides the gain
scaling of $1/k_g$ as depicted in Fig. 5.5.

5.3.4 Comparator

The comparator for the single-bit conversion is shown in Fig. 5.8. The regenerative comparator consists of an input differential-pair loaded by a negative resistance
due to the positive feedback of the cross-coupled transistors Plp and Pln. There
are two modes of operation, the mode of comparison (tracking) and the mode of
latching (regeneration) which are determined by the high-frequency clock-signal of
2.56 GHz [45].

During the mode of comparison both switches S_{track} are closed together with the
switch $\overline{S_{latch}}$. The latter switch needs to be closed to break the positive feedback of
the p-type latch (Plp and Pln). The gain of the differential-pair is sufficiently high to
cause a regeneration action of the previous decision. Switch S_{latch} is kept open, thus,
no current can flow through the n-type latch (Nlp and Nln). Closing switch S_{latch}
introduces the latch-phase whereas a small voltage-difference at the differential-pair

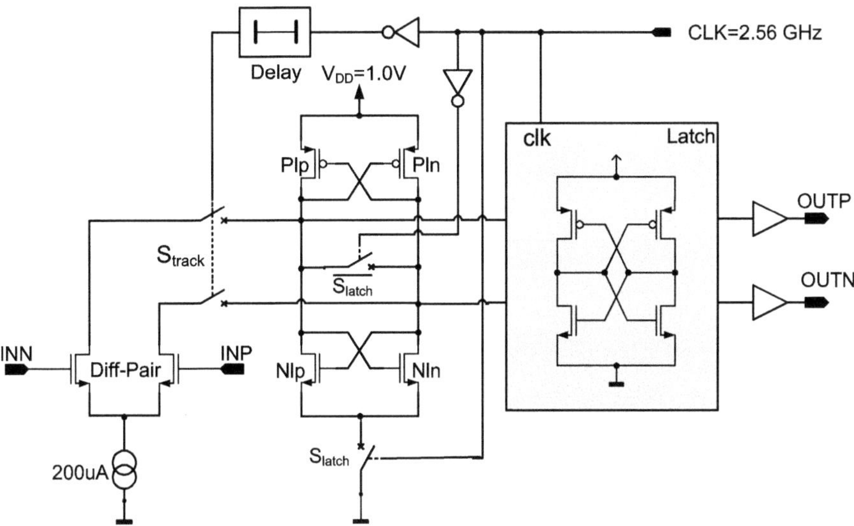

Fig. 5.8 Comparator for PWM-loop

output defines the initial condition. Next, the switch $\overline{S_{latch}}$ is opened to enable the dynamic-latch composed of both, the n-type and p-type latch. After a certain delay, both switches S_{track} are opened to decouple the input in order to avoid any possible kick-back mechanism. The output of the dynamic-latch is applied to a succeeding dynamic-latch for latching the taken decision (positive or negative) of the comparator. Before the next comparison phase is started, the succeeded dynamic-latch is decoupled from the comparator to latch its decision for further signal processing.

5.3.5 D/A-Converter in Feedback-Path

In [31] a novel architecture is presented for a multi-bit continuous-time delta-sigma modulator that replaces the conventional internal flash quantizer by a self-oscillating Pulse-Width Modulator. The main drawback of this technique is the need of an extra multi-bit D/A-converter with a large number of levels. This poses several implementation issues such as the need of a mismatch-shaping circuit to overcome the linearity limitations of the D/A-converter in the delta-sigma feedback loop. Nevertheless, this approach is specially suited to low-voltage technologies due to the absence of several comparators used in a ladder-configuration for comprising a flash-converter and may permit a significant reduction in power and area of existing data converters.

This work uses the architecture as shown in Fig. 5.5. In the outer-loop (DAC_2) the lowpass filter $H_{lp}(s)$ reduces the spectrally-shaped quantization noise of the single-bit data-stream. Furthermore, this lowpass filter acts as a tracking function for the

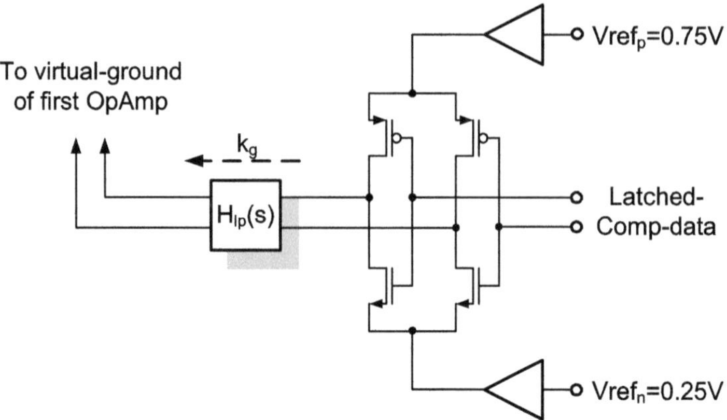

Fig. 5.9 DAC in feedback path

feedback signal, since it behaves similar to a lossy-integrator due to the pole placed at a low frequency. This tracking function leads to an amplification of the feedback signal, thereby, the feedback signal gets amplified by the weight-factor k_g, as discussed in the architecture section. Any impairments in the delta-sigma feedback path (DAC_2) deteriorates the overall performance since the feedback signal is subtracted from the input signal and any occurring error is present in the digitized output signal. Taking this issue into account, special care was taken during the design process for the D/A-converter as well as for the filter $H_{lp}(s)$. Beneficially, a single-bit DAC is required only revealing inherent linearity, as shown in Fig. 5.9. Furthermore, a passive RC-filter is sufficient for implementing $H_{lp}(s)$ which contains a real pole slightly above 20 MHz. Thanks to the provided 65 nm-technology, linear poly-resistors and linear inter-metal-capacitors are available for achieving sufficient linearity. On one hand, the noise contribution is constraining the RC-filter design only, on the other hand, smaller resistor values require larger capacitor sizes.

The latched pseudo-differential CMOS-signal is applied to the single-bit voltage-based D/A-converter comprised by two simple inverter stages, as shown in Fig. 5.9. Both inverters are supplied by two reference buffers without global feedback, as shown in Fig. 4.6, achieving a fast settling-time. The area described by the settling-transient of both reference buffers is crucial since the filtered DAC-output signal is directly subtracted in a continuous-time manner. Hence, the feedback signal is equal to the time-integral of the current-flow into the CT-loopfilter. Due to the dimensioning of the total A/D-converter (k_g, ROSR, COSR) one PWM-period is sampled $8\times$ by the clock-signal. Furthermore, it is very likely that the occurring PWM-pulses are not shorter than $3 \cdot 1/f_{clk}$. Nevertheless, the pulse-width modulated feedback-signal acts similar to a non-return-to-zero-DAC in conventional CT delta-sigma modulators [24].

Table 5.4 Full-scale voltage of WLAN A/D-converter

ω_{lp} [MHz]	K_c [MHz]	k_g [1]	g_{lp} [1]	V_{ref2} [V]	V_{FS} [Vpeak]
20	1280	6.7	1.52	0.5	0.164

5.3.6 Overall Circuit of A/D-Converter

Figure 5.10 depicts a schematic of the overall A/D-Converter for WLAN. The loopfilter comprises three OpAmps only revealing a power-drain of 4.3 mW in total. The time-encoding quantizer (TEQ) is built with passive filters only. The clock-frequency of 2.56 GHz triggers the comparator and the latch within the self-oscillating loop. Careful design and layout of the regenerative-comparator combined with the analog-latch results in a power consumption of typically 1.2 mW. The generated signal $Data_o$ and the complementary signal $\overline{Data_o}$ contains the modulated limit-cycle by the loopfilter output signal as well as the noise-shaped quantization error. The self-oscillating PWM-loop generates $f_{osc} = 320$ MHz containing the spectrum of the input signal beside some undesired spectral components. The modulated duty-cycle is discrete-in-time having a resolution of $1/2.56$ GHz $= 390$ ps. The remaining building blocks run at a lower rate of the limit-cycle frequency which helps to keep the total power drain small. Taking account for the RC-spreading due to process variations, all relevant capacitors are programmable by 3-bits covering a range of ± 50 %.

Generally, the full-scale signal of a delta-sigma converter is defined by the maximum amplitude of the feedback signal. The proposed solution uses a lowpass filter in the feedback path resulting in a tracking behavior of the input signal. In other words, the feedback path contains an integrator-function with a gain of g_{lp}. The voltage-DAC_2, the reference-voltage V_{ref2} and the gain g_{lp} define the full-scale voltage V_{FS} of the A/D-converter (5.15). Table 5.4 shows the values for calculating the full-scale voltage V_{FS} for a fully-differential sinusoidal wave whereas a reference voltage V_{ref2} is provided

$$g_{lp} = \frac{K_c}{k_g \cdot \omega_{lp}}, \tag{5.14}$$

$$V_{FS} = \frac{V_{ref2}}{2 \cdot g_{lp}}. \tag{5.15}$$

A small signal AC-model was derived to simulate the Noise Transfer Function (NTF) and Signal Transfer Function (STF) of the PWM-converter. The comparator was replaced by a simple gain-stage, whereas a gain of 35 was required for an AC-simulation to achieve a gain of one[7] at the limit-cycle frequency for the self-oscillating PWM-loop. The delay of the comparator together with the analog-latch

---

[7]Barkhausen's criteria for a self-oscillating loop: loop-gain ≥ 1 and phase-shift $= n \cdot 360°$, $n = 0, 1, 2, 3, \dots$.

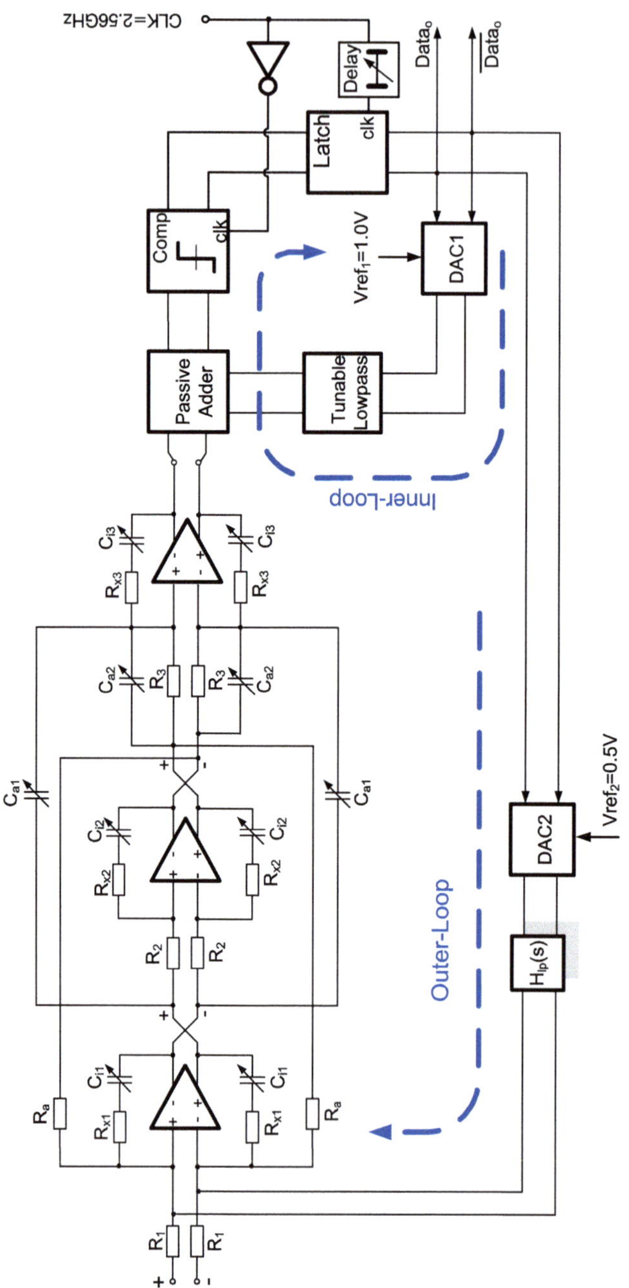

Fig. 5.10 Schematic of PWM A/D-converter

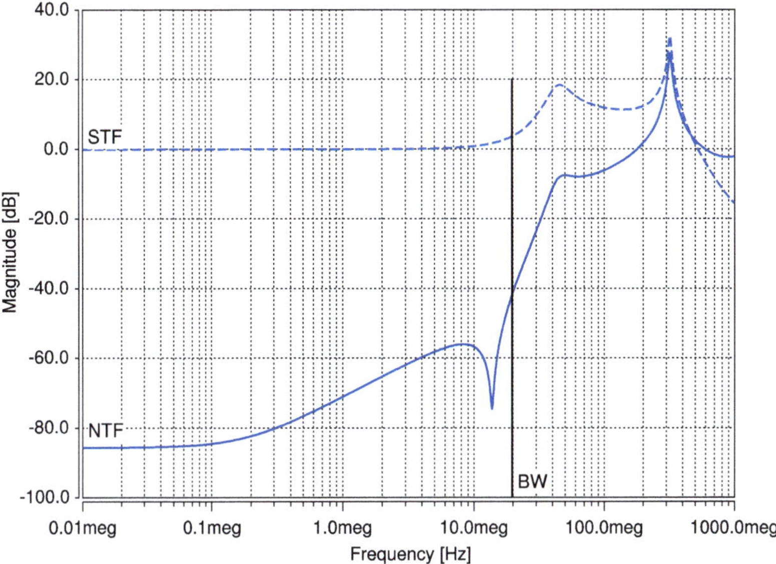

Fig. 5.11 Simulated NTF and STF of WLAN A/D-converter

is modeled by the Laplace-transform of a delay-cell (5.16). The delay is equal to one clock-period of $T_{clk} = 1/2.56$ GHz $= 390$ ps. The D/A-converter is described by the Laplace-transform of a non-return-to-zero DAC (5.17) as described in [24]. The feedback signal during idle-case (small-signal behavior) shows a mean-value equal to 50 % for the duty-cycle:

$$H_{del}(s) = e^{-sT_s}, \tag{5.16}$$

$$H_{DAC}(s) = \frac{1 - e^{-sT_s}}{sT_s}. \tag{5.17}$$

Figure 5.11 shows the simulated NTF (solid line) and STF (dashed line) of the PWM-converter. The NTF damps the quantization error more than 80 dB at low frequencies, whereas at the upper-end of the analog-bandwidth the conjugate-complex zeros provide a loop-gain between 60 dB and 40 dB. Thanks to the conservative loopfilter design, the NTF flattens out around 50 MHz without peaking. The peaking at 320 MHz is caused by the limit-cycle generated in the pulse-width modulator. The STF is flat at lower frequencies, whereby a slight amplification of 3 dB occurs at the analog bandwidth corner ($BW = 20$ MHz). Peaking up to 18 dB occurs at 45 MHz due to the feedforward loopfilter approach. The generated limit-cycle can be seen at 320 MHz before the STF starts to roll-off. This means that any undesired tones or signals are amplified towards the output of the modulator in the out-of-band frequency range. Nevertheless, this fact is not an issue, since for the targeted WLAN solution a high-order lowpass filter is placed in front of the converter to suppress the tones of the undesired adjacent channels after down-mixing [5, 49].

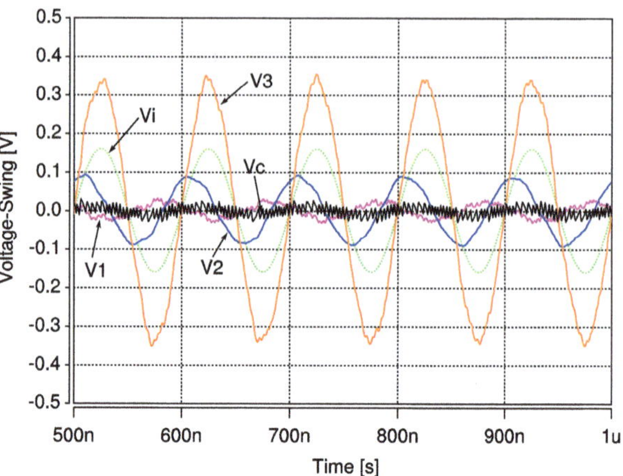

Fig. 5.12 Simulated voltage swing of state variables

Another characteristic of the feedforward loopfilter is that the input signal is mostly bypassed and the quantization noise is processed by the modulator. Thus, the signal swing at the output of the integrators is reduced which relaxes the output swing requirement for the OpAmps. Figure 5.12 depicts the signal swing of the state variables at the output of every OpAmp derived from a transient simulation. A full-scale signal (Vi) with a frequency of 10 MHz was applied at the input of the loopfilter. The signal swing at the output of the first (V1), second (V2) and third OpAmp (V3) is approximately 40 mVp, 100 mVp and 350 mVp, respectively. The signal at the input of the comparator (Vc) is superimposed by the scaled $(1/k_g)$ voltage (V3) and the generated limit-cycle revealing zero-crossings within every cycle. The limit-cycle is a triangular-waveform which results from $H_{lc}(s)$ imposing a integrating-function of a rectangular-waveform. Thanks to the feedforward loopfilter approach, the required signal swing can be kept quite small which is in favor with a low-supply voltage of 1 Volts at the expense of none intrinsic anti-aliasing filtering.

5.4 Measurement Results

The A/D-converter was implemented in a 65 nm digital CMOS technology *without* using any process options such as a thick-gate-oxide device. The standard copper metal-stack offers 4-thin metal-layers, one medium thick and one fat metal-layer for power supply routing. The provided core-device needs a supply voltage of 1.0 V ± 10 % to guarantee a lifetime up to 15 years maintaining sufficiently small parameter drifts. The gate-oxide thickness is below 2 nm, thus, the occurring leakage current through the thin gate-oxide cannot be ignored anymore during the design procedure. The linear capacitors were implemented as advanced sandwich devices reducing manufacturing complexity as compared to MIM-capacitors at the expense of a less dense layout, since no devices can be placed underneath this type

Fig. 5.13 Layout of WLAN A/D-converter

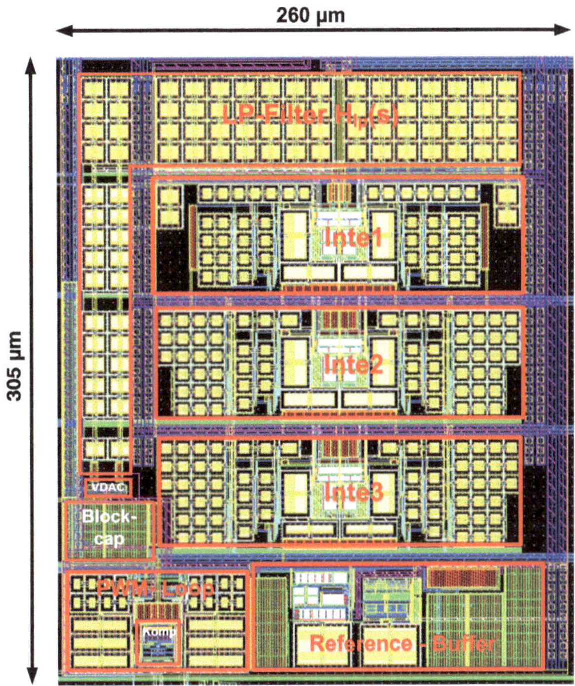

of sandwich capacitor. The advanced sandwich devices employ both the latheral- and vertical-capacitances between proper arranged metal-stripes of all 4-thin metal-layers to achieve a high specific capacitance per unit area. The single poly-layer serves as the basis for processing of the linear resistor.

Figure 5.13 depicts the layout of the A/D-converter-core showing an attractively small size of $260\,\mu m \times 305\,\mu m = 0.079\,mm^2$ including the voltage reference buffer. Most of the silicon area needs to be spent for arrays of linear unit-capacitors of the loopfilter whereas the consumed area by resistors and active MOS-devices play a minor role. The capacitor arrays are implemented in common-centroid topology for achieving good matching. The three integrators are stacked on top of each other enabling a straight signal-flow from the top-side towards the bottom-side. Thanks to the simplicity of the single-bit voltage-based D/A-converter (see Fig. 5.9), the required chip-area is negligible small. The occurring voltage spikes during signal transition are damped by decouple-capacitors placed nearby the D/A-converter. The lowpass filter in the delta-sigma feedback loop realizes a real pole slightly above 20 MHz in order to damp the out-of-band spectral components such as the shaped quantization noise and the limit-cycle. In order to keep the contributed thermal noise of the resistors sufficiently below the 10 bit noise floor, a relatively large portion of the whole A/D-converter needs to be reserved for the lowpass filter.

The capacitors of the filter $H_{lp}(s)$ (see Fig. 5.10) are distributed along the left corner serving as a barrier between the high-frequency part (TEQ or PWM-loop) and the sensitive input of the loopfilter. Any error caused by the TEQ will be third-

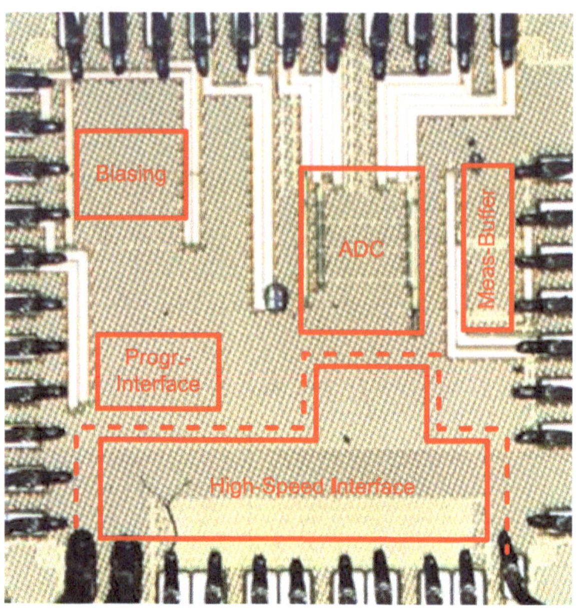

Fig. 5.14 Photograph of test-chip

order noise shaped, thus, a high thermal-noise contribution can be tolerated leading to an attractively small area for the whole PWM-loop, as shown in the lower-left corner. This allows to go for a very compact layout keeping parasitics small which helps to get little delays within the high-speed circuitry clocked at 2.56 GHz. Care was taken that the digital switching circuits do not interfere with the functionality of the analog building blocks. Therefore, guard rings and substrate contacts are provided that shield the digital building-blocks from the CT filters. Only a single V_{DD}- and V_{SS}-rail was chosen for supplying the whole A/D-converter since in the targeted product application a further separation of power domains could not be afforded.

The evaluation of the prototype needs to be done very carefully due to practical constraints in measurement. The testchip comprises in addition to the A/D-converter a reference buffer for the D/A-converter in the outer-loop as well as a bandgap, voltage-to-current converters, a programming interface, a measurement buffer and a high-speed interface. A chip micrography is depicted in Fig. 5.14. The bandgap circuit provides a stable on-chip reference-voltage of 500 mV which is lowpass filtered with an external capacitor ($C = 220$ nF) to reduce the contribution of noise. The filtered bandgap voltage is applied to voltage-to-current converters providing bias-currents to all OpAmps as well as a reference-current for the voltage-reference buffer. All generated currents can be tuned within a ± 50 % range to have the possibility for trading between power consumption and performance, if needed. The implemented digital serial-interface enables the communication between a personal-computer and the testchip. Thus, the digital interface provides access to all programming on-chip registers for controlling the loopfilter coefficients, the programming of the limit-cycle frequency by setting the delay within the PWM-loop. Furthermore, the settings for all relevant bias-currents, the choice for the input-signal to the mea-

Fig. 5.15 Photograph of test-board

surement buffer and the setting for the high-speed interface can be controlled via the digital interface.

The measurement buffer serves as a debug feature and can be used for tapping the voltage right at the input of the loop-filter or at the output of the first, second or third integrator in order to check the signal amplitude and the noise- and distortion-figures. The high-speed interface acts as transmitter for the single-bit data captured right at the input of the D/A-converter in the outer-loop. The minimum theoretical pulse-width results in 400 ps whereas the limit-cycle gets $8\times$ oversampled revealing a limit-cycle period of typical 3.2 ns. The information of the single-bit data-stream can be found in the quantized duty-cycle which carries the input signal as well as the shaped quantization noise and the limit-cycle.

Further on, the receiver part of the high-speed interface senses differentially the externally applied 2.5 GHz clock-signal. The sensed clock is amplified and distributed as a current-mode-logic (CML) signal in order to keep any possible voltage spikes as small as possible on the common supply rails. Exactly at the placement of the comparator, the CML-clock is converted back into a CMOS clock-signal for triggering the sampling process. A guard-ring was routed around the high-speed interface to reduce any possible coupling to the sensitive parts of the testchip (depicted as dotted line).

5.4.1 Measurement Setup

Figure 5.15 shows a photograph of the designed testboard. Several power supply connectors are placed on the right hand side. A 3.3 V supply is used for the biasing circuit including the bandgap as well as the voltage to current converters. The A/D-converter is supplied by a separated 1.0 V supply to measure exactly the current consumption. The high-frequency interface has its own connectors avoiding any crosstalk over the supply rails. The digital programming interface runs over the 9-pin connector placed on the opposite side. On the top side of the board the input signal can be applied in a differential manner via two SMA-connectors. A single-ended input is supported as well by employing a transformer for the single-ended to fully-differential conversion (the transformer is not shown in Fig. 5.15).

The high-frequency interface is placed on the bottom side of the board. The required 2.5 GHz clock-signal is applied fully-differential via two SMA-connectors complying with a 50 Ω matching impedance. The converter's output data can be captured at both SMA-connectors which are driven fully-differential by an on-chip high-frequency buffer. A triangular shape of the lower edge allows a short on-board routing of high-frequency signals. The testchip is mounted directly on the board to avoid any parasitics related to a package. The bondwires are designed as short as possible to result in a sufficiently small bondwire inductance in order to prevent ringing on the supply rails. A four layer plain-copper-board (PCB) was considered to be adequate. Both middle layers are used for a V_{DD}- and a V_{SS}-plane and the upper and lower plane are used for signal routing. Special care was taken for avoiding any crosstalk between the high-frequency interface running at 2.5 GHz and the device-under-test, the A/D-converter. As a consequence, the V_{SS}-plane is split into two parts, the lower part supplies the high-speed interface whereas the remaining part provides a general ground plane. This splitting helps to reduce a possible coupling from the high-frequency interface to the sensitive building blocks such as the reference buffer or the biasing circuits. Both parts of the V_{SS}-plane are connected by several equally distributed coils to keep the same DC-level on the whole board. Several check points were arranged to measure the bandgap voltage, the generated bias- and reference-currents as well as the on-chip common-mode voltage of all OpAmps. All decoupling capacitors for buffering of supply voltages are soldered on the bottom side of the PCB.

Figure 5.16 illustrates the measurement setup for the evaluation of the PWM-ADC. A signal-generator provides a single-ended sinusoidal waveform which is applied to a transformer for making available a fully-differential input signal to the A/D-converter. A passive lowpass filter is inserted at the output for achieving sufficient low distortion by damping the second and third harmonic component accordingly. The input signal range covers a frequency band from 100 kHz up to 20 MHz and amplitudes between 160 μVp and 160 mVp. The common-mode voltage can be adjusted by a voltage divider having 0.5 V as a target value. The signal generator is synchronized to the 2.5 GHz clock-signal by a 10 MHz synchronization signal in order to avoid skirts around the desired tone in the FFT. A high-frequency sinusoidal signal-generator serves as a clock-generator for providing the 2.5 GHz clock-signal. Special care needs to be taken for the phase-noise contribution by the clock-source equipment. The sampling jitter caused by the clock-signal has been considered to result in a minimum SNR of 72 dB. Taking (2.62) yields for a signal-frequency of 20 MHz a long-term jitter requirement of smaller than 2.8 ps. The available digital pattern-generators on the current market do not provide such a jitter performance for a 2.5 GHz clock-signal, thus, a sinusoidal signal-generator was considered to be adequate.

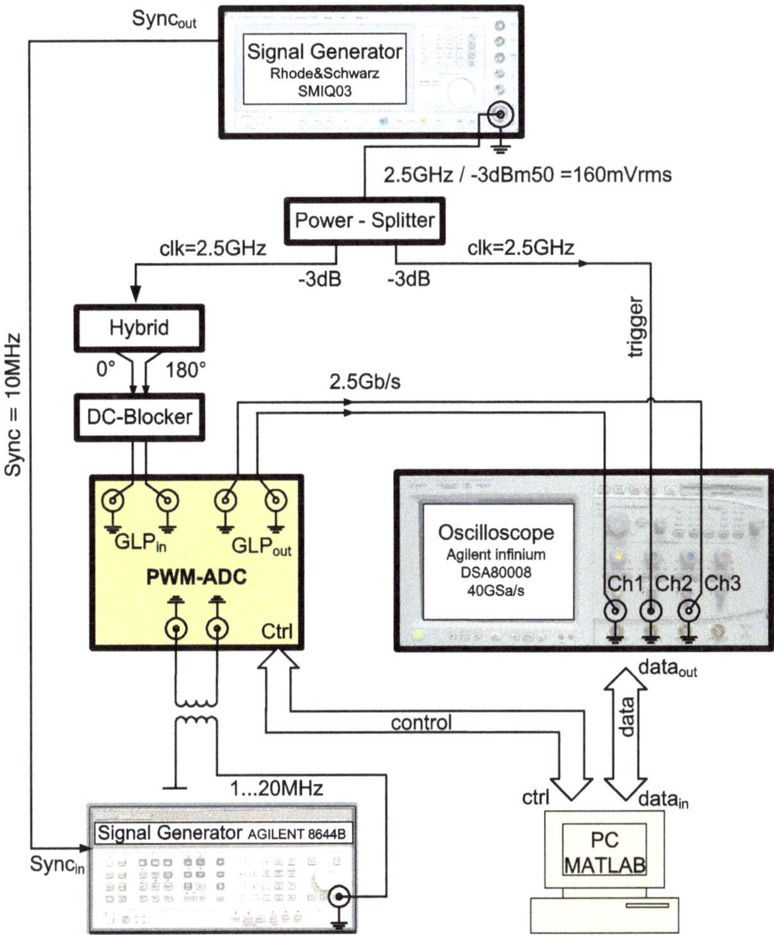

Fig. 5.16 Measurement setup for WLAN A/D-converter

The single-ended 2.5 GHz sinusoidal signal is coherently split into two signals by
an external power-splitter. One signal provides the clocking for the A/D-converter,
the other one is used for triggering a digital oscilloscope. The former signal is further applied to an external hybrid-circuit which generates two output signals with
the same frequency and a phase-shift of 180°. After removing the DC-component
by two DC-blockers, the fully-differential 2.5 GHz clock-signal is applied to the
high-speed interface of the test-chip. The latched comparator's data output stream
is connected to the output buffers of the high-speed digital interface. The buffers'
output data come out as fully-differential signals employing current-mode logic.
A digital oscilloscope is used for capturing the differential signal by subtracting
channel-1 from channel-3. The output data-stream is synchronized on-chip to the
clock-signal, hence, the oscilloscope needs to be triggered by the 2.5 GHz signal-

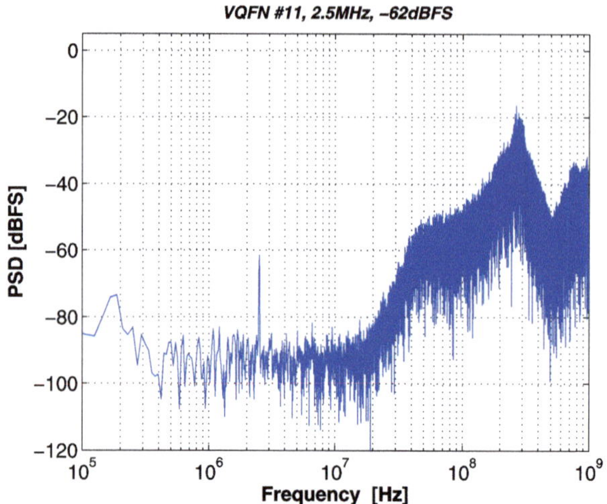

Fig. 5.17 PSD plot of PWM-ADC for −62 dBFS input signal

generator for ensuring a proper data capturing. The captured data are stored in the oscilloscope's data memory and are readout by a PC running a particular Matlab© program. The whole testchip is controlled by the PC via the digital programming interface.

5.4.2 Measurement of PSD-Plots

The FFT result for a −62 dBFS input signal is depicted in Fig. 5.17 in which the spectrum was processed with 120k data points. (0 dBFS refers to a sine wave with 320 mVpp.) The data-stream was captured right at the digital input of the D/A-converter without any digital lowpass filtering. The band-of-interest up to 20 MHz is dominated by white noise. The total sum of both resistors at the input and the resistors of the lowpass filter $H_{lp}(s)$ contribute thermal noise up to −63 dBFS. Above 20 MHz the third-order noise-shaping function can be seen by a slope of +60 dB/decade. The noise-shaping settles around 50 MHz without showing any overshoot which demonstrates the robustness of the delta-sigma loop. The position of the limit-cycle is below 300 MHz due to parasitic components of relevant interconnection metal lines. It turned out, that the parasitic extraction tool within the provided design flow showed too optimistic results yielding a deviation of 15 %. Hence, the trimming-range of the whole PWM-loop was shifted towards lower frequencies.

Figure 5.18 depicts a PSD-plot with a tone at −42 dBFS. The position and the shape of the limit-cycle remains unchanged. For larger input amplitudes the limit-cycle becomes wider due to the stronger modulation of the duty-cycle, as shown in Fig. 5.19 for an input signal with an amplitude of −12 dBFS. Figure 5.20 illustrates the PSD-plot with a tone 3 dB below full-scale. The limit-cycle gets wider

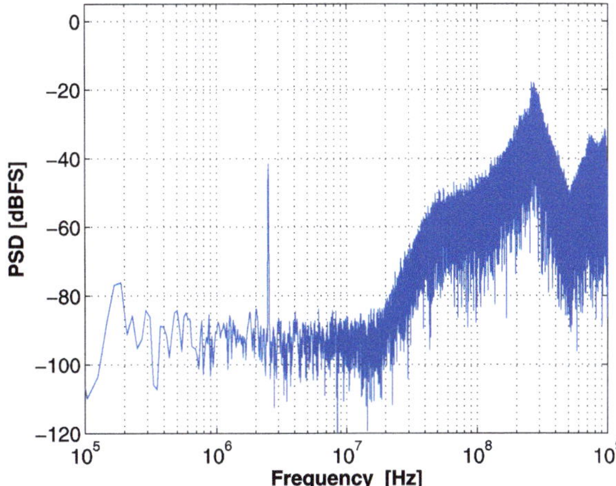

Fig. 5.18 PSD plot of PWM-ADC for −42 dBFS input signal

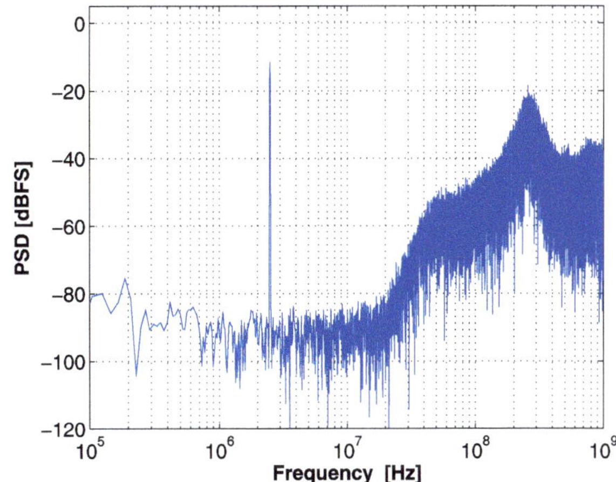

Fig. 5.19 PSD plot of PWM-ADC for −12 dBFS input signal

and shifted towards lower frequencies. A second harmonic component is visible. The root-cause was found in mismatch between both single-ended voltage-reference buffers of the outer-loop. The mismatch between the positive and negative reference buffer caused a difference between the rising- and falling-edge at the output of the D/A-converter of the pulse-width modulated binary signal. At full-scale, a third harmonic component comes into place, as shown in Fig. 5.21. Distortion attracts attention with −68 dBc for the 2nd harmonic and −66 dBc for the 3rd harmonic. At 20 MHz, the third-order shaped-noise becomes dominant. The shape of the limit-cycle broadens further due to a strong modulation of the duty-cycle between 20 % and 80 %.

Figures 5.22 and 5.23 show the measured Signal-to-Noise Ratio (SNR) and Signal-to-Noise and Distortion Ratio (SNDR) versus the normalized input signal $\frac{V_{in}}{V_{REF}}$, respectively. The frequency of the input signal was at 2.5 MHz and the data-

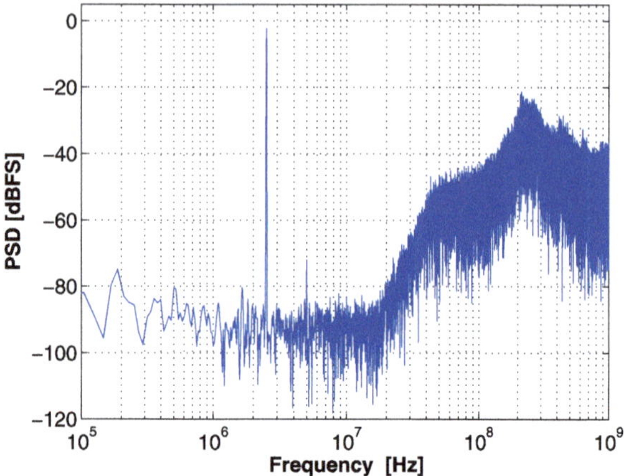

Fig. 5.20 PSD plot of PWM-ADC for -3 dBFS input signal

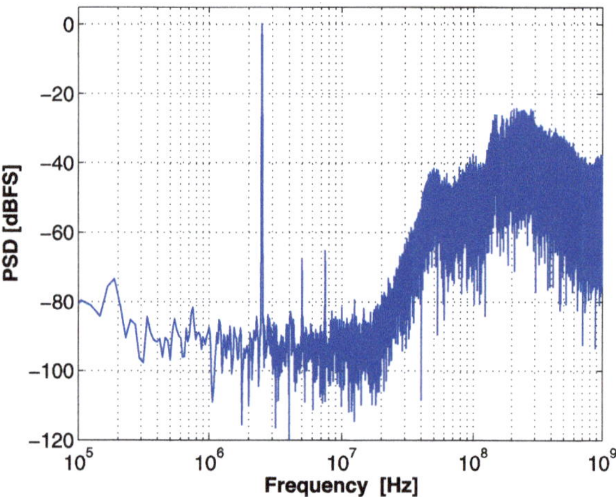

Fig. 5.21 PSD plot of PWM-ADC for 0 dBFS input signal

stream was captured right at the input of the D/A-converter. The postprocessing was done twice in Matlab, first by taking the raw data for the FFT and second by applying the raw data to an equalizer implementing the transfer function $H_{lp}(z)$ as shown in Fig. 5.5. The transfer function of the used equalizer is equal to $H_{lp}(s)$, that means, it matches the frequency response of the lowpass filter in the outer-loop. Applying the equalizer improves the SNR and SNDR by more than 1 dB over the whole input range. Up to -2 dBFS the noise energy is dominant over the energy of the harmonics. The 3rd-order harmonic component is caused by the OpAmps of the loopfilter, whereas the 2nd-order harmonic component originates in mismatch between the rising-time and falling-time of the edges of the binary D/A-converter. The measured distortion components (HD2 and HD3) are depicted in Fig. 5.24.

The measured Signal Transfer Function (STF) of the PWM-ADC is illustrated in Fig. 5.25. Omitting the equalizer reveals the impact of the lowpass filter in the outer-loop. The feedback signals is damped at higher frequencies leading to less

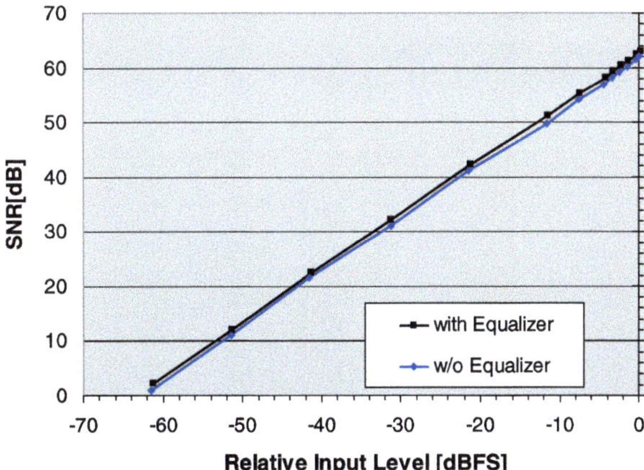

Fig. 5.22 Measured signal-to-noise ratio

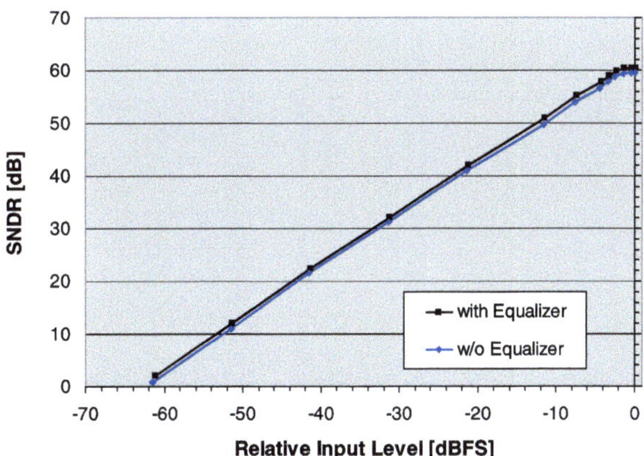

Fig. 5.23 Measured signal-to-noise and distortion ratio

feedback causing an increased signal amplitude. The equalizer compensates the rise of the STF at higher inband frequencies by a proper lowpass filtering. The remaining ripple shows the behavior of a conventional continuous-time delta-sigma modulator employing a similar loopfilter. Furthermore, the SNR for a 15 MHz input tone is depicted in Fig. 5.26, whereas the equalizer was applied as well.[8] Due to the rising behavior of the STF, the SNR-curve shows a peaking 2 dB below full-scale, since the delta-sigma modulator is close to its overload condition for tones at higher input frequencies.

Figure 5.27 illustrates a Discrete Multi-Tone (DMT) signal which is common for WLAN applications. Each modulated tone carries a certain number of bits for data transmission. The resulting peak-to-rms ratio (CREST-factor) of the DMT-signal varies due to the instant bit-allocation. The resulting CREST-factor can go up to 4 (12 dB) challenging the noise and linearity requirements. The Missing-Tone Power

---

[8]Note: For an input tone at 15 MHz the SNDR-curve is equal to the SNR-curve, since any occurring harmonic components are placed above the band of interest ($BW = 20$ MHz).

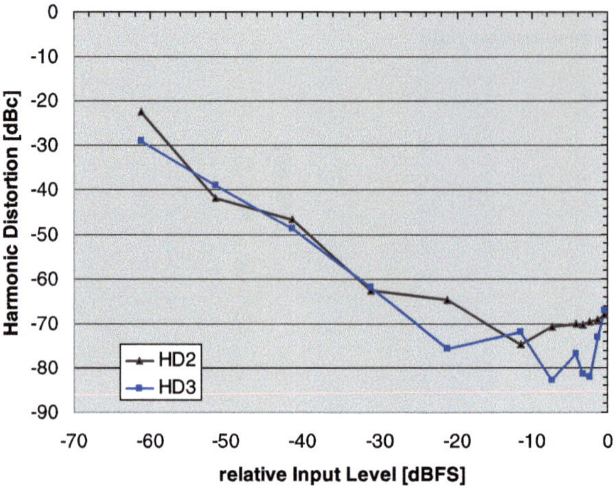

Fig. 5.24 Measured distortion components of PWM-ADC

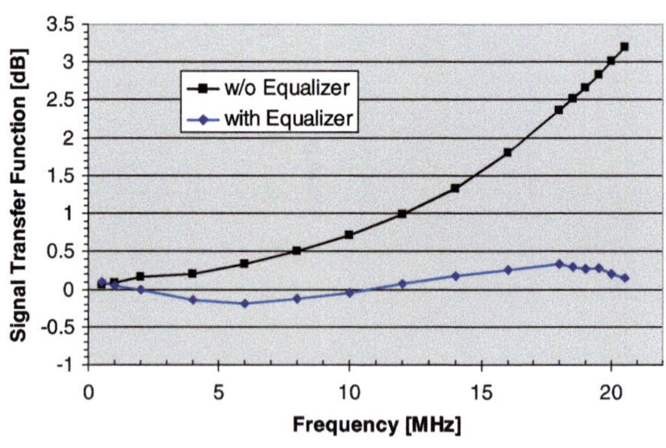

Fig. 5.25 Measured signal transfer function of PWM-ADC

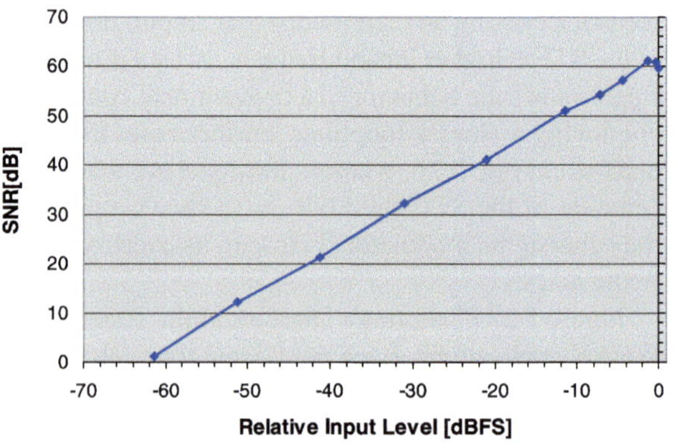

Fig. 5.26 Measured signal-to-noise ratio for 15 MHz input tone

Ratio (MTPR) evaluates the overall linearity of a WLAN-system. Hereby, some of the DMT-tones are omitted, as the 10th, 20th, 30th, 40th, 50th and 59th tone in Fig. 5.27 (see arrows). The distance between the level of DMT-tones and the

Fig. 5.27 Measured missing-tone power ratio for *CREST* = 4

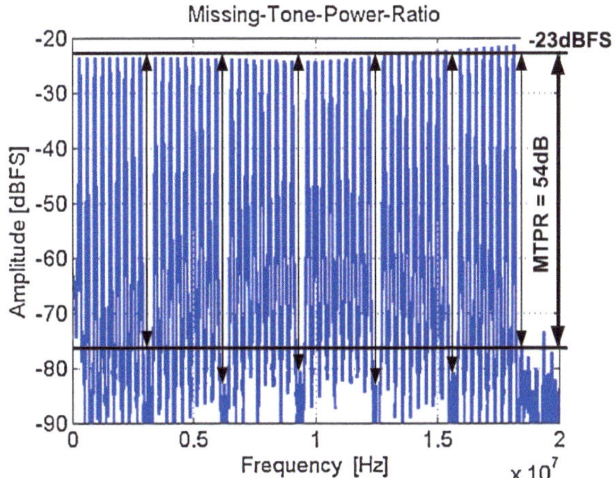

Table 5.5 Measured MTPR of PWM A/D-converter

Tone number	10th-tone	20th-tone	30th-tone	40th-tone	50th-tone	59th-tone
Frequency [MHz]	3.125	6.250	9.375	12.500	15.625	18.4375
MTPR [dB]	54	64	61	62	60	57

sum of all occurring intermodulation products at the omitted tones is larger than 54 dB. Thus, the achieved linearity and noise contribution is sufficient for maintaining maximum data-rate. Table 5.5 summarizes the five-times measured and averaged MTPR's at each omitted tone.

The TEQ is embodied by the inner-loop of the A/D-converter, which replaces a multi-bit quantizer (see Fig. 5.10). The evaluation of the inner-loop only is shown in Fig. 5.28 by a PSD-plot of the captured output data. The loopfilter was bypassed in order to apply directly the input signal to the TEQ. Furthermore, the outer-loop was disabled, thus, the noise-shaping feature is switched off. The shown spectra reveals a limit-cycle slightly above 300 MHz. The applied input level is quite small (-45 dBFS) which results in a small modulation of the duty cycle. Hence, the shape of the limit-cycle tone is quite close to that of a single discrete tone. The applied input tone is visible at 2.5 MHz whereas no distortion components can be observed above the given noise floor. The dynamic-range plot of the inner-loop is depicted in Fig. 5.29. Above an input level of -45 dB the distortion components become dominant over the noise-power within an bandwidth up to 20 MHz. The measured peak-SNR and peak-SNDR results in 40 dB and 35 dB, respectively. The TEQ is supposed to replace a 4-bit quantizer, thus, the measured resolution of the TEQ fulfills the required resolution. Table 5.6 summarizes some of the important measurement data.

Fig. 5.28 PSD plot of inner-loop for −45 dBFS input signal

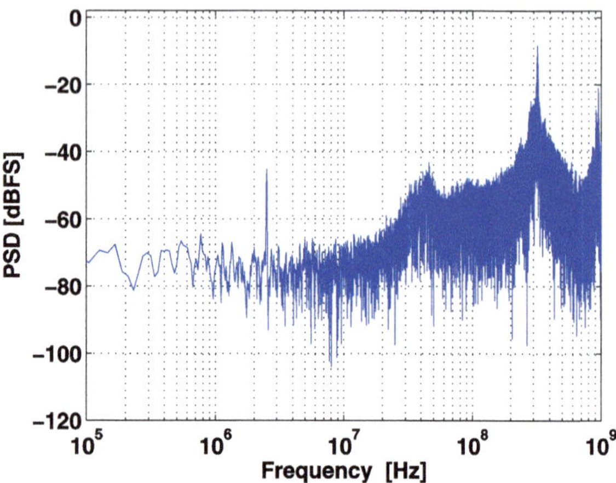

Fig. 5.29 Measured signal-to-noise ratio of inner-loop (TEQ)

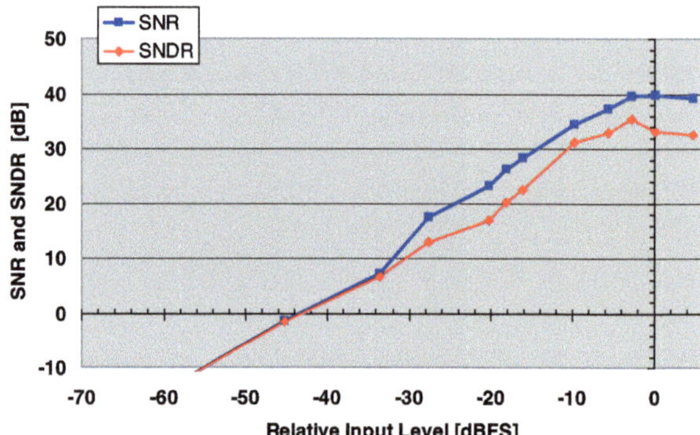

5.5 Digital Decimation Filter

A decimation-filter is usually placed after the delta-sigma modulator in order to make the converted signal usable for further digital signal-processing. A decimation-filter reduces the data-rate and increases the bit-widths accordingly by keeping the achieved SNR. In general, the decimated data stream after the digital filter is usually the same as for a Nyquist-rate converter. Thus, after the decimation filter we could get a quite similar data-stream as compared to an equivalent Flash A/D converter. The proposed architecture provides a single-bit data-stream at a very high rate of 2.56 GS/s. The high data-rate might cause the need for a complex decimation filter along with high power consumption for driving the digital circuitry. The risk of high power drain can be avoided by choosing a proper filter architecture, as discussed below.

As a proposal, a first draft concept of the digital filter is shown in Fig. 5.30. The output of the WLAN-ADC modulator is a 1-bit serial data-stream running at a high data-rate of 2.56 GBit/s. This high data-rate cannot be processed by using standard-logic digital design methods. Therefore, the data-rate has to be reduced to a lower

Table 5.6 Measured performance summary of WLAN A/D-converter

Signal bandwidth	20 MHz
Virtual sampling frequency	640 MHz
Clocking frequency	2.5 GHz
Limit-cycle frequency	300 MHz
Reference voltage	0.5 V (diff.)
Dynamic range (DR)	63 dB
Peak SNR	63 dB
Peak SNDR	61 dB
Total power of ADC $V_{DD} = 1.0$ V	7.0 mW
Technology	65 nm 1.0 V digital CMOS 6ML
FOM (dynamic-range)	0.15 pJ/conv
FOM (peak SNR)	0.15 pJ/conv
FOM (peak SNDR)	0.19 pJ/conv

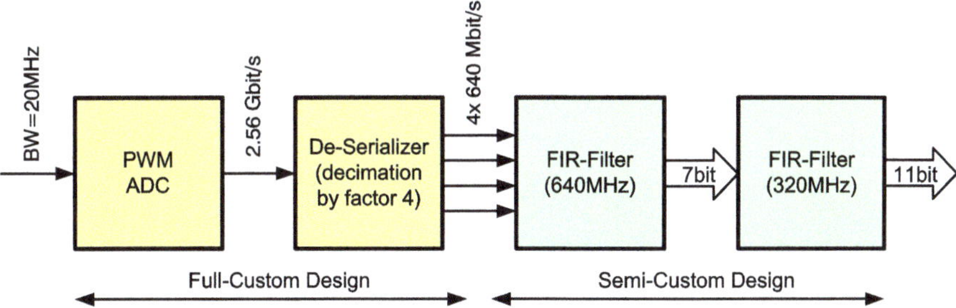

Fig. 5.30 Proposal for decimation-filter path for WLAN-ADC

data-rate feasible for digital design tools. Switching to a lower frequency as soon as possible results in a power efficient implementation. This decimation could be done by a de-serializer circuit that can be implemented as full-custom circuit block. As an example, the approach of a polyphase filter can be chosen to save input shift-registers by using a multiplexer instead. The decimation factor should be 4 to reduce the clock frequency to about 640 MHz which might be feasible for a digital standard cell design in 65 nm CMOS technology. A possible implementation is depicted in Fig. 5.30, which is described briefly. Following the full-custom 1-to-4 decimation block the PWM digital filter consists of two Finite Impulse Response (FIR) filters with integer coefficients. The first FIR-filter has a 4-bit input data coming from the de-serializer and a 7-bit output data. The filter is running at 640 MHz clock frequency. The following filter stage makes an additional decimation by a factor of 2 and has an output data of 11 bit widths, running at a clock frequency of 320 MHz. A full custom digital design was carried out on schematic level as a proof of concept. Pipelining of the whole digital data stream was necessary in order to avoid any setup- and hold-violations. A static timing analysis (STA) was performed to verify a robust timing. The design was simulated on SPICE level as well and the overall power drain from a single 1.0 V supply resulted below 1.5 mW implemented in a 65 nm CMOS technology.

5.6 Conclusion

This chapter showed the operating principle and experimental results of a new architecture for a continuous-time delta-sigma modulator. The proposed modulator does not require a multi-bit quantizer nor a mismatch shaping D/A-converter to produce a multi-bit noise-shaped output signal. In contrast to conventional $\Delta\Sigma$-modulators, the coarse quantizing of the loopfilter output is accomplished by a time-encoding technique similar to a Pulse-Width Modulation (PWM) producing a binary signal. That means, a PWM employs the high-speed features of nanometer technologies to overcome the limitations of reduced signal swing by the very low supply voltage of 1.0 V. The offered nanoscale high-speed devices enable the exchange of *amplitude-resolution by time-resolution*.

Despite the high clock-rate $f_{clk} = 2.5$ GHz of the comparator output, the loopfilter and the single-bit DAC of the proposed modulator have the same speed requirements ($f_s = 640$ MHz) as those of the equivalent conventional multi-bit ($N = 4$) delta-sigma modulator. On the other hand, in the proposed modulator there are no multi-bit converters used (neither analog-to-digital nor digital-to-analog). Therefore, for a given set of specifications, the proposed PWM Delta-Sigma modulator is proven to consume less power and area than its equivalent conventional multi-bit modulator. The complexity of a multi-bit flash quantizer in a multi-bit Delta-Sigma modulator is much bigger as compared to the complexity of a single regenerative comparator. Utilizing this fact, the output of the proposed modulator is a binary signal only and the D/A-converter in its feedback path is single-bit as well. Furthermore, single-bit-DACs are inherently linear so that no trimming or DEM technique is required. In other words, this new technique offers a multi-bit-performance at a low level of complexity of a single-bit implementation. The single-bit stream containing the PWM-signal is used to generate the feedback for the delta-sigma loop as well as the free running limit-cycle loop. Thanks to the sampling process *within* both loops, the generated data are discrete in time, thus, the data-stream can be easily applied to conventional digital signal-processing. It is worth mentioning that the proposed technique of pulse-width modulation can be adapted for a wide range of resolutions similar to conventional delta-sigma modulation. Furthermore, the proposed PWM-circuitry can be scaled in a straight-forward manner for smaller feature-size technologies such as 40 nm or 32 nm.

The new architecture for the A/D-converter was implemented in a 65 nm digital CMOS technology without requiring any process options to achieve 10 ENOB's within a bandwidth of 20 MHz. The measured total power drain from a single 1.0 V supply resulted in 7 mW including the $\Delta\Sigma$-modulator along with a reference buffer. The calculated FOM according to (2.7) results in 0.15 pJ/conversion-step. Furthermore, the total area of the $\Delta\Sigma$-modulator consumes 0.079 mm^2 of silicon area only.

Both achieved figures demonstrate the high potential of this new technique for A/D-conversion. This A/D-converter was presented at ESSCIRC 2010 conference [8] and published in IEEE Journal of Solid-State Circuits [7].

Chapter 6
Conclusions

In general, A/D-converters are divided into two broad categories: *Nyquist-rate* converters and *over-sampling* converters. These different converter classes typically offer different compromises between ADC resolution and output sampling rate. Nyquist-rate converters are those that operate at a minimum sampling frequency necessary to capture all the information about the entire input bandwidth. Three of the most popular Nyquist-rate converter architectures are SAR (successive approximation register), flash and pipeline ADCs. Basically, Nyquist-rate converters are used in open-loop configuration without any global feedback. A constant input during the conversion process usually is provided in each stage by a sample-and-hold (SHA) circuit. The first-stage SHA must maintain the accuracy of the overall ADC at the full sampling rate, requiring the circuit to settle within a single clock period. In contrast, oversampled delta-sigma A/D-converters do not require highly accurate settling circuits as compared to Nyquiste-rate converters and thus avoid the increased power drain and need for high-performance drivers in high-resolution applications. This advantage comes mainly from the global feedback structure providing noise-shaping in the frequency domain at the expense of a limited band-of-interest defined by the Over-Sampling Ratio (OSR). Nevertheless, CT delta-sigma modulators may also include significant anti-aliasing filtering, reducing or eliminating the need for an additional anti-aliasing-filter. Finally, CT delta-sigma technology is well-suited for migration to future CMOS processes.

High-performance CMOS A/D-converters suitable for System on Chip (SoC)-integration in a digital transceiver face constraints due to noisy supply-rails, on-chip cross-talk, wide temperature ranges from $-40\,°C$ to $+120\,°C$ and on top of that all process variations occurring during fabrication. All impairments need to be taken into consideration in the design phase in order to fulfill the product specifications.

6.1 Research Work Overview

In this book the research work is presented about the design of broadband A/D-converters for product development in the field of communications. Oversampling

R. Gaggl, *Delta-Sigma A/D-Converters*, Springer Series in Advanced Microelectronics 39, 127
DOI 10.1007/978-3-642-34543-2_6, © Springer-Verlag Berlin Heidelberg 2013

A/D-converters employing the principle of delta-sigma modulation were taken into consideration. In particular, two designs for products for the ADSL-Central-Office applications were discussed in a 0.18 μm and 0.13 μm technology node. Further on, one novel A/D-converter design for WLAN-application was shown in a 65 nm technology node. The main focus was put on low-power consumption and very small chip-area to result in a competitive product design for the world market. At the same time, attention was paid to get sufficient design robustness for achieving almost no yield-loss in high volume production.

In chapter one a quick overview is given about the system requirements on ADSL- and WLAN-solutions. The main requirements on the A/D-converter are specified in terms of bandwidth, resolution and linearity.

The second chapter deals with the main limitations on oversampled A/D-converters. Beside the basics of A/D-conversion, the main operating principles and theoretical performances of delta-sigma modulators are shown. The differences between a switched-capacitor and a continuous-time implementations are addressed. The impact of circuit nonidealities on the overall performance are discussed and simplified models are presented for system-level simulations. An introduction is given to the principle of asynchronous delta-sigma modulation. Finally, the clock-jitter issue and its impact to the total SNR is pointed out.

A single-loop second-order switched-capacitor solution for ADSL-CO applications is presented in chapter three. A multi-bit feedback was chosen together with an OSR of 96 to achieve 14 ENOBs over a bandwidth of 276 kHz. An implementation in a 0.18 μm-technology is shown employing a novel dynamic-biasing technique for power saving. Considerations on architectural level are given. The main issues in circuit design are pointed out. Attention was paid to the design of a dynamically biased two-stage amplifier used for a switched-capacitor loopfilter. The targeted performance was demonstrated by measurement results of several prototypes under all operating conditions.

In chapter four a novel feedforward architecture is introduced for achieving a low-power consumption along with an attractive small chip-area. A second-order switched-capacitor modulator with an reference buffer in open-loop configuration was implemented in a 0.13 μm-technology for an ADSL-CO application. A 3 bit quantizer clocked at 105 MHz was considered to maintain 14 ENOBs. As an option, 13 ENOBs can be achieved over a bandwidth of 1.1 MHz showing a flat in-band noise density. Beside architectural considerations on a feedforward SC-solution, relevant issues on circuit level are addressed such as the design of a canonic-integrator and the passive SC-adder. The measurement results as well as more than several million pieces sold manifest the reliability performance of this delta-sigma converter design.

A third-order continuous-time delta-sigma solution is presented in chapter five aiming for a WLAN application. A coarse multi-bit quantizer is replaced by a novel time-encoding technique employing the resolution in time instead of resolution in amplitude. This approach overcomes the limited signal swing issue due to reduced supply voltages in nanoscale technologies. The novel A/D-converter architecture can be scaled in a straight-forward manner for smaller feature-size technologies such as 40 nm or 32 nm. Design considerations on system-level were carried out for the implementation of a time-encoding quantizer inside a delta-sigma loop. This approach made it possible to replace a multi-bit quantizer together with a feedback DAC by a binary circuit. It was shown that the loopfilter of the delta-sigma modulator may be seized to spectrally shape both the time quantization error and the interpolation error. The new architecture makes use of a pulse-width modulator inside a noise-shaping loop resulting in a rather simple circuit similar to a single-bit implementation but revealing multi-bit performance. The pulse-width modulator was implemented by a passive circuit avoiding any active stages to keep the power drain low. Finally, the design of a third order CT delta-sigma modulator is discussed to show the potential of the proposed technique which demonstrates the circuit simplification. The evaluation of a test-chip in a 65 nm technology showed a performance of 10 ENOBs over a bandwidth of 20 MHz. Comprehensive measurement results and discussions on the evaluation issues along with the very low-power drain and consumed chip-area show the attractiveness of the new approach.

6.2 Figure-of-Merit Comparison

In recent years, delta-sigma A/D-converters have got more and more attention in communication products beside the Nyquist-type converters such as pipeline-ADC, flash-ADC, subranging-ADC and successive-approximation-register (SAR). Driven by the advances in deep-submicron technologies as well as the introduction of multi-bit architectures and continuous-time delta-sigma ADCs, world wide industry has made enormous efforts to employ delta-sigma modulators with analog bandwidths ranging from a few kHz (voice) up to a few multiple of 10 MHz (WLAN, VDSL). The main two advantages of noise-shaping converters are the possible low power consumption and the required small chip-area at the expense of a needed OSR giving an upper limit to the analog bandwidth.

An introduction of a Figure Of Merit (FOM) is needed in order to compare different designs for the same application. This means, a single number is calculated on the basis of relevant performance parameters demonstrating the efficiency of a given design solution. Thus, a FOM is defined by the ratio of all required resources over the achieved performance. The investment is given for A/D-converter designs by the required power-drain as well as the consumed chip-area, as shown by (6.1) (power-drain) and (6.2) (chip-area).

The power-FOM (figure-of-merit) is defined in (6.1):

$$Power\text{-}FOM = \frac{P}{2 \cdot BW \cdot 2^{(SNR-1.76)/6}} \quad [\text{J/conversion-step}]. \tag{6.1}$$

Furthermore, an area-FOM (figure-of-merit) is introduced by (6.2):

$$Area\text{-}FOM = \frac{Area}{2 \cdot BW \cdot 2^{(SNR-1.76)/6}} \quad [\text{mm}^2 \text{ s/conversion-step}]. \tag{6.2}$$

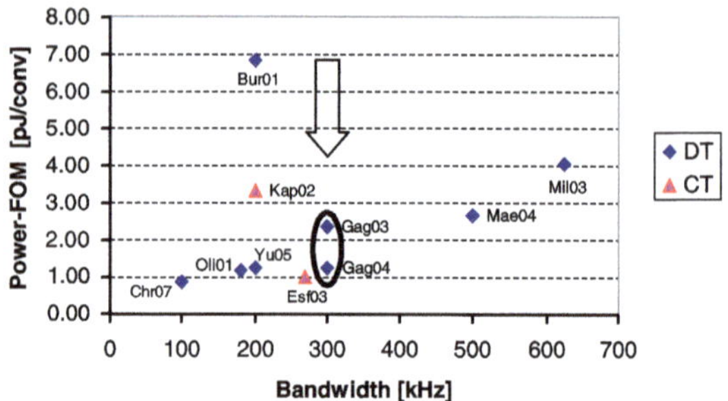

Fig. 6.1 Power-FOM as a function of signal bandwidth for ADSL-CO

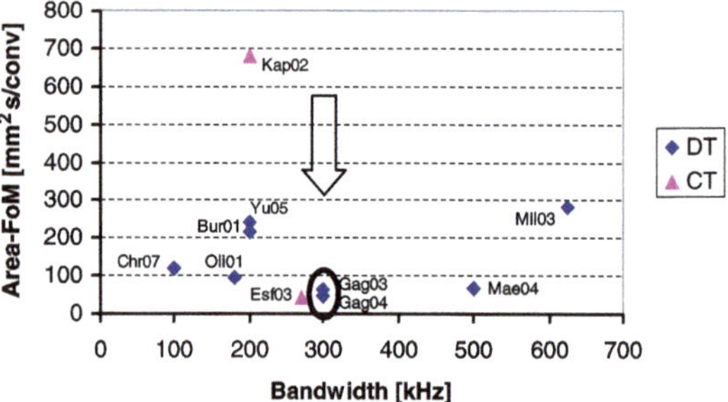

Fig. 6.2 Area-FOM as a function of signal bandwidth for ADSL-CO

A survey of papers in the delta-sigma area was carried out for ADSL-CO specifications, that means $BW \approx 300$ kHz and $ENOB \approx 13.5$ bit, and for WLAN applications, that means $BW \approx 10 \ldots 20$ MHz and $ENOB \approx 10$ bit.

The published state of the art delta-sigma converter designs for ADSL Central-Office specifications are listed in Table 6.1. Recently published Continuous Time (CT) and Discrete Time (DT) modulators with an analog bandwidth between 100 kHz and 600 kHz are considered, as shown in Fig. 6.1. Only the power-drain and the consumed chip-area of the converter-core is considered since the numbers for the reference circuits are not always published.

The A/D-converters presented in chapter three and four are given in [47] = ([Gag03]) and [48] = ([Gag04]), respectively. The achieved power-FOM keeps up with the best ones whereas the needed chip-area is fairly small, as illustrated by the area-FOM in Fig. 6.2. This was achieved by exploiting the high-speed devices by going for an high OSR to avoid a MASH-structure. Note, both developed solutions were optimized for a bandwidth of 300 kHz whereas some of the other converters offer a multi-standard solution. Thanks to the high OSR, Ref. [48] supports the bandwidth of 1.1 MHz as well revealing a reduction in DR of 1 bit only. Both DT-modulators, Chr07 and Oli01 (see Table 6.1), reveal a slightly better power-FOM at the expense of a MASH topology leading to a larger silicon area. Due to noise-leakage, MASH converters are more prone to circuit impairments which increases the risk for achieving a robust yield in high-volume production.

Both CT-modulators, Kap02 and ESF03 (see Table 6.1), demonstrate the low-power feature of a continuous-time loopfilter. Nevertheless, the increased jitter sensitivity of the feedback signal requires a very low-jitter clock. Thus, more power-

Table 6.1 Overview of recently published $\Delta\Sigma$-modulators for ADSL-CO

Publication	Reference	Techn. [nm]	Loopfilter [–]	f_s [MHz]	BW [kHz]	SNR [dB]	SNDR [dB]	Power [mW]	Area [mm^2]	P-FOM [pJ/conv]	A-FOM [mm^2 s/conv]
[22]	Yu05	90	DT	20	200	74	72	2.1	0.4	1.25	237
[68]	Chr07	130	DT	26	100	86	85	2.9	0.4	0.86	119
[48]	**Gag04**	**130**	**DT**	**105**	**300**	**82**	**80**	**8.0**	**0.15**	**1.26**	**46**
[47]	**Gag03**	**180**	**DT**	**53**	**300**	**82**	**78**	**15**	**0.2**	**2.36**	**63**
[34]	Mil03	180	DT	23	625	77	77	30	2.08	4.03	280
[66]	Bur01	250	DT	104	200	74	71	11.5	0.36	6.83	214
[40]	Oli01	350	DT	13	180	83	82	5.0	0.40	1.17	93
[21]	Mae04	600	DT	24	500	82	81	28	0.70	2.64	66
[33]	Kap02	180	CT	20	200	64	63	1.8	0.36	0.09	679
[13]	Esf03	250	CT	13	270	80	78	4.6	0.19	1.01	43

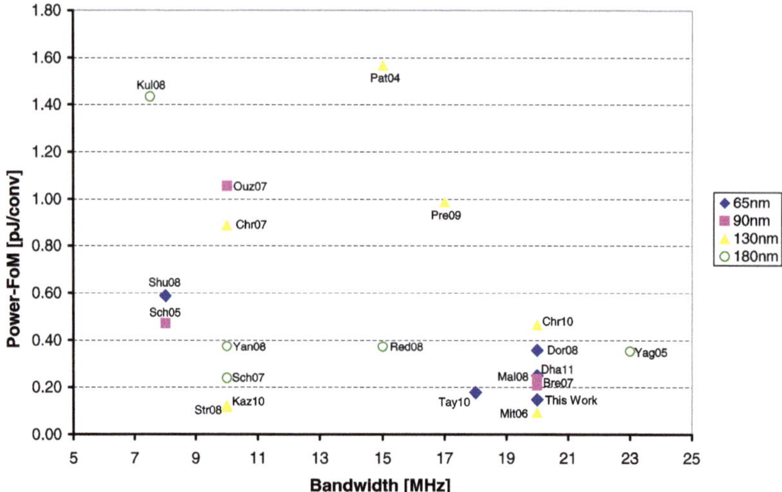

Fig. 6.3 Power-FOM as a function of signal bandwidth for WLAN

drain needs to be spent for the clock-synthesizer and the whole clock-tree for a multi-channel system.

In [47] the modulator was optimized on circuit-level by introducing a dynamic-biasing technique. Although a single-loop solution was given preference as compared to [4], the FOM could still be enhanced. On the contrary, in [48] the ADC was already improved on architectural level by employing a feedforward approach. This example demonstrates how beneficial it is to tackle the trade-off between power and performance already during architectural considerations. It is worth mentioning that for lower bandwidths the DT-solutions can still keep up with CT-ones in spite of their higher speed requirement for all OpAmps. Scaling of delta-sigma modulators over advanced technology-nodes does not work without proper adaptations on the overall architecture to overcome the limitations of high-speed devices suffering from smaller gain per transistor-stage.

The published state of the art continuous-time delta-sigma converter designs for WLAN-specifications are listed in Table 6.2. Recently published modulators with an analog bandwidth higher than 7 MHz are considered. Figure 6.3 depicts the according power-FOMs with the reported performance. It can be seen that nowadays most broadband leading-edge designs achieve a FOM much lower than 1 pJ/conversion-step. The four different technology-nodes (180 nm, 130 nm, 90 nm and 65 nm) are pointed out by different symbols, as shown in legend of Fig. 6.3.

A clear trend over the evolution of technology nodes cannot be seen in the different FOMs. All technology nodes are spread over the whole range of analog bandwidths by covering FOMs up to 1.6 pJ/conversion-step. It seems, as the right choice on architectural level as well as a careful circuit design results in an attractive FOM. Thus, analog-design does not shrink automatically over the different technology nodes. Innovation seems to be the key-factor for achieving attractive FOMs. The

Table 6.2 Overview of recently published $\Delta\Sigma$-modulators for WLAN

Publication	Reference	Techn. [nm]	f_s [MHz]	BW [MHz]	SNR [dB]	SNDR [dB]	Power [mW]	Area [mm²]	P-FOM [pJ/conv]	A-FOM [mm² s/conv]
this work	**this work**	**65**	**2500**	**20**	**63**	**61**	**7**	**0.079**	**0.15**	**1.5**
[30]	Dor08	65	640	20	62	60	15	0.18	0.36	4.3
[78]	Shu08	65	256	8	76	70	50	0.50	0.59	5.9
[70]	Dha11	65	250	20	62	60	11	0.15	0.25	3.6
[16]	Tay10	65	1152	18	70	67	17	0.07	0.18	0.7
[41]	Mal08	90	420	20	72	70	28	1.00	0.21	7.5
[29]	Bre07	90	340	20	71	69	28	0.25	0.24	2.1
[59]	Ouz07	90	400	10	52	50	7	0.20	1.06	30.2
[51]	Sch05	90	1000	8	64	63	10	0.35	0.47	16.6
[28]	Kaz10	110	300	10	68	63	5	0.32	0.13	7.6
[68]	Chr07	130	240	10	63	63	21	0.40	0.89	16.9
[15]	Mit06	130	640	20	76	74	20	1.20	0.09	5.7
[36]	Str08	130	900	10	86	72	40	0.42	0.12	1.3
[61]	Pat04	130	300	15	65	64	70	1.20	1.57	26.9
[11]	Pre09	130	1623	17	59	59	25	0.1	0.99	4.0
[67]	Chr10	130	400	20	67	64	35	0.27	0.47	3.6
[26]	Red08	180	300	15	67	63	21	1.00	0.37	2.6
[74]	Yan08	180	640	10	84	82	100	0.37	0.70	31.3
[55]	Kul08	180	240	8	71	67	64	1.40	1.43	7.4
[39]	Yag05	180	276	23	70	69	43	0.90	0.35	7.8
[52]	Sch07	180	640	10	66	66	8	0.26	0.24	17.8

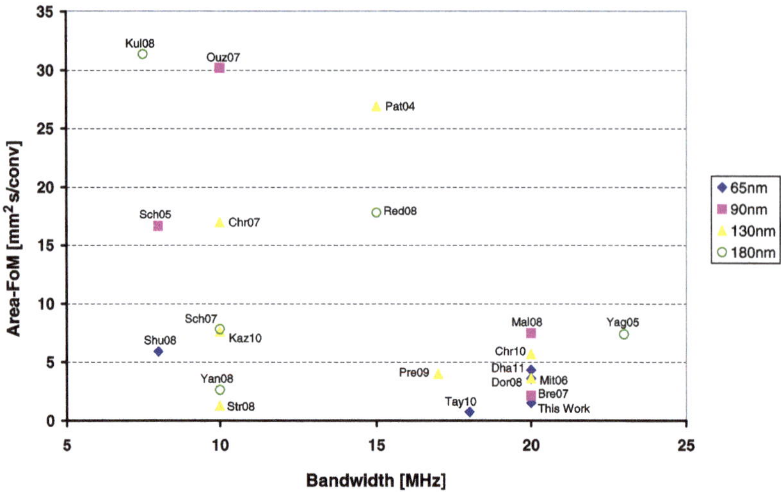

Fig. 6.4 Area-FOM as a function of signal bandwidth for WLAN

features of new technologies such as high-speed devices need to be exploited in order to improve the converter designs. The reduced supply voltage limits the possible signal-swing, thus, for resulting in the same SNR the noise-floor needs to be lowered accordingly. This straightforward approach asks for smaller values for resistor causing stronger loading for the amplifiers which increases the power-drain. Furthermore, for any given RC-time-constant the capacitors need to be increased in size to compensate for the smaller resistor value. One possible way for solving this drawback is to exchange the amplitude-resolution by time-resolution. In other words, the high-speed feature of nanoscale technologies can be employed to overcome the issues caused by reduced signal-swing in A/D-converter design.

The consumed silicon area of 0.079 mm^2 for this work (see chapter five) is remarkably small, as shown in Table 6.2 and illustrated by the comparison of all area-FOMs in Fig. 6.4. The solution of [36] = Stro08 employs a VCO-based quantizer to replace a 5-bit flash converter inside a first-order delta-sigma converter. This approach results in a good SNR, however, it suffers from the nonlinear transfer-curve of a voltage-controlled oscillator (VCO). Thus, the measured peak-SNDR deteriorates by 12 dB as compared to the peak-SNR. This drawback is not visible in Figs. 6.3 and 6.4, since both refer to the peak-SNR, as defined in (6.1) and (6.2). Another VCO-based approach is published in [16] showing a rather small chip size for a continuous-time delta-sigma modulator for analog-to-digital conversion because it consists mostly of digital circuitry implemented in 65 nm CMOS. It is a voltage-controlled ring oscillator based design with new digital background calibration and self-canceling dither techniques applied to enhance performance. Unlike conventional delta-sigma modulators, it does not contain analog integrators, feedback DACs, comparators, or reference voltages. Therefore, it uses less area than comparable conventional delta-sigma modulators, and the architecture is well-suited to IC processes optimized for fast digital circuitry.

Generally speaking, scaling of delta-sigma modulators over advanced technology-nodes does not work without proper adaptations on the overall architecture to overcome the limitations of high-speed devices suffering from smaller gain and swing per transistor-stage.

6.3 Outlook on Future Trends in Delta-Sigma Design

This book deals with the design of delta-sigma modulators in the field of communications. The targeted applications are ADSL and WLAN, whereas the former one requires high-resolution converters and the latter one needs high-speed data-conversion. Highly integrated transceiver solutions in modern CMOS technologies for wired and wireless communication systems require high-performance and low-power ADCs that consume only small silicon area and run from a single supply voltage. As shown in the pipeline and SAR arena, the delta-sigma community will develop new architectures and combinations of known topologies in order to achieve more bandwidth along with high efficiency.

As an example for extending the state-of-the-art for the WLAN converter two steps of improvements are briefly pointed out. Firstly, a time-to-digital-converter (TDC) could be used as described in [63] instead of the presented pulse-width-modulator (PWM) in Chap. 5. Although a single-bit pulse-width modulator could be further exploited for more performance, a TDC might be an option, since the required high clock-rate might become impracticable for some product solutions. In spite of the fact that a TDC shows a higher effort on circuit-level in terms of power-drain and complexity, a TDC inside a delta-sigma loop will allow to go for a higher bandwidth along with higher resolution. Of course, a TDC with its inherent signal delay will increase the excess loop delay which deteriorates the loop stability. Secondly, an optimization on circuit-level of the WLAN-ADC might be of interest, since more than 2/3 of the total power goes into the loopfilter. This fact gives motivation for a new OpAmp design in 65 nm, either by more aggressive design such as conditionally stable OpAmps or by changing the type of integrators. In the thesis of [50] it was concluded that $g_m C$-integrators are more power efficient than RC-integrators, if a rather moderate accuracy is demanded (p. 120). The reported numbers based on simulation results in 90 nm CMOS [51] and 180 nm CMOS [52] showing a slightly worse power-FOM than this work, as depicted in Table 6.2. Although it is not exactly defined what a moderate accuracy means for a 65 nm technology, it might be worth going for a $g_m C$-integrator design.

The continuous progress in A/D-converter designs involves exploiting the well known trade-offs between analog bandwidth, power efficiency and signal-to-noise ratio (SNR). The high speed of nanometer CMOS technologies has improved the performance of classical Successive Approximation Register ADC (SAR) and pipeline ADC architectures. SAR converters are known for their simplicity and low area, but they are limited due to the power for the needed reference buffers and known linearity issues of the preceding sample and hold stage. A time interleaved approach can overcome these limitations by parallel data processing at the expense of increased power and chip area.

Pipeline A/D-converters are limited by switch nonlinearity and settling requirements in low voltage technologies, due to the nature of a global open-loop topology. The switched-capacitor implementation for a pipeline A/D-converter requires highly accurate settling transients to achieve the target performance. The need for proper settling transients demands fairly high speed OTAs along with high DC-gain. Both requirements of high speed and high DC-gain ask for challenging OTA designs resulting in high power drain and larger chip area. As a pipeline converter can be considered as an open-loop topology, it cannot make use of any kind of error shaping as it is possible in a delta-sigma modulator. Recent publications have shown new and improved design combinations of pipelined SAR converters. These new architectures have shown a power efficiency in the range of 10 fJ/conv. An example for the combination of two converter types can be found in Ref. [6]. A fully dynamic, two-times interleaved pipelined SAR ADC is presented that achieves a power-FoM of 10 fJ/conversion-step with 9.5 ENOB at a sampling speed as high as 250 MS/s. Another example for a combination of a SAR and a noise-shaping approach is shown in reference [20]. This paper introduces a low-OSR noise-shaping SAR ADC that leverages noise-shaping to increase the resolution of a conventional SAR ADC. The converter uses an 8-bit capacitor DAC and achieves an ENOB of 10 bits over a bandwidth of 11 MHz with an OSR of 4. The power-FoM for this converter is 35.8 fJ/conv.

The ongoing trend for wide-band, power-efficient continuous-time $\Delta\Sigma$-modulators has led to various implementations, which commonly share the usage of multi-bit quantization and low oversampling ratio. Digitally assisted approaches have been introduced in $\Delta\Sigma$-modulators recently, such as in [44]. This paper shows a full finite GBW compensation of the loop-filter in combination with a digitally estimated and corrected feedback DAC, enabling low-speed amplifiers with low parasitic input load. The power-FoM of 88 fJ/conv for a 25 MHz bandwidth is achieved by the combination of robust finite GBW compensation, a 4-bit quantizer, a 3rd-order loopfilter, digitally error estimated analog DAC linearization, and a power-aware circuit implementation.

In our days, the achievable signal bandwidth for oversampled A/D-converters have already exceeded the 100 MHz range, as demonstrated at ISSCC 2011 conference and published in JSSC [37]. This paper presents a wideband 3rd continuous-time delta-sigma ADC with a loopfilter topology that absorbs the pole caused by the input capacitance of its 4-bit quantizer and also compensates for the excess delay caused by the quantizer's latency. The bandwidth is 125 MHz and the quantizer is clocked at 4 GHz. The ADC was implemented in 45 nm CMOS and achieves 70 dB dynamic-range, while dissipating 260 mW from 1.1 V and 1.8 V supply. The straight-forward implementation of conventional high-speed $\Delta\Sigma$-modulators at clock-rates in the multi-Gb/s speeds has been further shown in [43] and [69] at ISSCC 2012 conference. Generally speaking, at multi-Gb/s speeds the maximum sampling rate is set by Excess Loop Delay (ELD) considerations. A shown in [43] a finite impulse response (FIR) DAC is used to reduce sensitivity to clock-jitter and to relax loopfilter linearity. A mostly analog path compensates the modulator for the delay introduced by the FIR DAC. The modulator employs a 4th-order loopfilter along with a single-bit comparator sampled at 3.6 GS/s. Measurements result in a 83 dB dynamic-range over a 36 MHz bandwidth and the ADC occupies 0.12 mm^2 in 90 nm CMOS. Dissipating 15 mW from a 1.2 V supply, it thereby achieves a power-FoM of 72.8 fJ/conv. For a wideband single-bit modulator clocking at high speeds (multi-GHz), the key challenge is modulator stability due to large ELD set by comparator speed limitations and integrator parasitic poles. The paper [69] describes circuit techniques to minimize ELD and a compensation scheme that ensures modulator stability given a single clock-period ELD. These techniques have enabled the design of a 3rd-order single-bit modulator clocked at 6 GHz in 45 nm CMOS. The measured SNDR in a 60 MHz bandwidth is 60 dB. The ADC power consumption is 20 mW resulting in a power-FoM of 190 fJ/conv, whereby it operates off of 1.8 V for the feedback DACs, 1.4 V for the analog supply and 1.4 V for the digital supply.

The ongoing bandwidth and SNR extension demands clock jitter towards 0.1 ps which challenges the PLL design and the whole clock distribution. On the other hand, continuous-time sigma-delta converters can benefit from nanometer CMOS technologies by new architectures that exchange amplitude by time resolution. These new ADC architectures can neatly compete in FoM and area with the latest SAR and pipeline converters and clearly show the potential to surpass them in terms of linearity, configurability and anti-aliasing features. An example for a voltage-controlled oscillator (VCO) based analog-to-digital conversion was presented at ISSCC 2012 conference [25]. A VCO-based analog-to-digital conversion presents an attractive means of implementing high-bandwidth oversampling ADCs. They exhibit inherent noise-shaping properties and can operate at low supply voltages and high sampling rates. However, usage of VCO-based ADCs has been limited due to their nonlinear voltage-to-frequency transfer characteristic. The VCO nonlinearity can be suppressed by embedding the VCO in a $\Delta\Sigma$-loop and by minimizing the input signal processed by the VCO.

Bibliography

1. A.A. Lazar, L.T. Toth, Time encoding and perfect recovery of bandlimited signals, in *Proceedings of ICASSP, International Conference on Acoustics, Speech and Signal Processing* (2003), pp. 709–712
2. A.M. Abo, P.R. Gray, A 1.5-V, 10-bit, 14.3-MS/s CMOS pipeline analog-to-digital converter. IEEE J. Solid-State Circuits **34**(5), 599–606 (1999)
3. A. Wiesbauer, J. Hauptmann, P. Laaser, Sigma delta converters in wireline communications, in *Proceedings of 11th Workshop on Advances in Analog Circuit Design (AACD)*, Spa, Belgium (2002)
4. A. Wiesbauer, H. Weinberger, M. Clara, J. Hauptmann, A 13.5-bit cost optimized multi-bit delta-sigma ADC for ADSL, in *Proceedings of the 25th ESSCIRC* (1999), pp. 82–88
5. B. Razavi, *RF Microelectronics* (Prentice Hall, New Jersey, 1998)
6. B. Verbruggen, M. Iriguchi, J. Craninckx, A 1.7 mW 11b 250 MS/s 2x interleaved fully dynamic pipelined SAR ADC in 40 nm digital CMOS, in *IEEE ISSCC Proceedings* (2012), pp. 466–467
7. E. Prefasi, S. Paton, L. Hernandez, A 7 mW 20 MHz BW time-encoding oversampling converter implemented in a 0.08 mm^2 65 nm CMOS circuit. IEEE J. Solid-State Circuits **46**(7), 1562–1574 (2011)
8. E. Prefasi, S. Paton, L. Hernandez, R. Gaggl, A. Wiesbauer, J. Hauptmann, A 0.08 mm^2, 7 mW time-encoding oversampling converter with 10 bits and 20 MHz BW in 65 nm CMOS, in *Proceedings of the 36th ESSCIRC* (2010), pp. 430–433
9. E. Roza, Analog to digital conversion via duty-cycle modulation. IEEE Trans. Circuits Syst. II **44**(11), 907–914 (1997)
10. E. Roza, Poly-phase sigma-delta modulation. IEEE Trans. Circuits Syst. II **44**(11), 915–923 (1997)
11. E. Prefasi, L. Hernandez, S. Paton, A. Wiesbauer, R. Gaggl, E. Pun, A 0.1 mm^2, wide bandwidth continuous-time $\Delta\Sigma$ ADC based on a time encoding quantizer in 0.13 µm CMOS. IEEE J. Solid-State Circuits **44**(10), 2745–2754 (2009)
12. F. Colodro, A. Torralba, M. Laguna, Continuous-time sigma-delta modulator with an embedded pulsewidth modulation. IEEE Trans. Circuits Syst. I **55**(3), 775–785 (2008)
13. F. Esfahani, Ph. Basedau, R. Ryter, R. Becker, A fourth order continuous-time complex sigma-delta ADC for low-IF GSM and EDGE receivers, in *Symposium on VLSI Circuits Digest of Technical Papers* (2003), pp. 75–78
14. G.M. Jin, F.O. Eynde, W. Sansen, A high-speed CMOS comparator with 8-b resolution. IEEE J. Solid-State Circuits **27**, 208–211 (1992)
15. G. Mitteregger, Ch. Ebner, St. Mechnig, Th. Blon, Ch. Holuige, E. Romani, A 20-mW 640-MHz CMOS continuous-time $\Sigma\Delta$ ADC with 20 MHz signal bandwidth, 80-dB dynamic-range and 12-bit ENOB. IEEE J. Solid-State Circuits **41**(12), 2641–2649 (2006)

R. Gaggl, *Delta-Sigma A/D-Converters*, Springer Series in Advanced Microelectronics 39, 139
DOI 10.1007/978-3-642-34543-2, © Springer-Verlag Berlin Heidelberg 2013

16. G. Taylor, I. Galton, A mostly-digital variable-rate continuous-time delta-sigma modulator ADC. IEEE J. Solid-State Circuits **45**(12), 2634–2646 (2010)

17. I. Fujimori, L. Longo, A. Hirapethian et al., A 90-dB SNR 2.5-MHz output-rate ADC using cascaded multibit delta-sigma modulation at 8x oversampling ratio. IEEE J. Solid-State Circuits **35**, 1066–1073 (2000)

18. I. Marocco, A noise model platform for discrete-time sigma-delta A/D converters in VLSI design. Technical report, Diploma thesis at Politecnico di Bari, 2001; diploma thesis of colaboration between Infineon and Politecnico di Bari, Tutor: R. Gaggl

19. J. Daniels, Asynchronous delta-sigma A/D-converters: digital demodulation using a time-to-digital converter. Technical report, Katholieke Universiteit Leuven, ESAT-MICAS, 2005; internal report of colaboration between Infineon and KU Leuven, ESAT-MICAS

20. J. Fredenburg, M. Flynn, A 90 MS/s 11 MHz bandwidth 62 dB SNDR noise-shaping SAR ADC, in *IEEE ISSCC Proceedings* (2012), pp. 468–469

21. J. Maeyer, P. Rombouts, L. Weyten, A double-sampling extended-counting ADC. IEEE J. Solid-State Circuits **39**(03), 411–418 (2004)

22. J. Yu, F. Maloberti, A low-power multi-bit $\Sigma\Delta$ modulator in 90-nm digital CMOS without DEM. IEEE J. Solid-State Circuits **40**(12), 2428–2436 (2005)

23. J.R. Higgins, R.L. Stens, *Sampling Theory in Fourier and Signal Analysis* (Oxford University Press, Oxford, 1996)

24. J.S. Cherry, W.M. Snelgrove, *Continuous-Time Delta-Sigma Modulators for High-Speed A/D Conversion* (Kluwer Academic, Dordrecht, 2000)

25. K. Reddy, S. Rao, R. Inti, B. Young, A. Elshazly, M. Talegaonkar, P.K. Hanumolu, A 16 mW 78 dB-SNDR 10 MHz-BW CT $\Delta\Sigma$ ADC using residue-cancelling VCO-based quantizer, in *IEEE ISSCC Proceedings* (2012), pp. 152–153

26. K. Reddy, Sh. Pavan, A 20.75 mW continuous-time $\Delta\Sigma$ modulator with 15 MHz bandwidth and 70 dB dynamic range, in *Proceedings of the 34th ESSCIRC* (2008), pp. 210–213

27. K. Vleugels, S. Rabii, B.A. Wooley, A 2.5 V broadband multi-bit sigma delta modulator with 95 db dynamic range, in *IEEE ISSCC Proceedings* (2001)

28. K. Matsukawa, Y. Mitani, M. Takayama, K. Obata, S. Dosho, A. Matsuzawa, A fifth-order continuous-time delta-sigma modulator with single-OpAmp resonator. IEEE J. Solid-State Circuits **45**(4), 697–706 (2010)

29. L. Breems, R. Rutten, R. Veldhoven, G. Weide, H. Termeer, A 56 mW CT quadrature cascaded $\Sigma\Delta$ modulator with 77 dB DR in a near zero-IF 20 MHz band, in *IEEE ISSCC Proceedings* (2007), pp. 238–239

30. L. Doerer, Continuous time $\Sigma\Delta$-3rd order multibit (4bit) for 802.11n applications (WLAN11n). Technical report, Infineon, Austria, Villach, 2008; internal report of Infineon Austria

31. L. Hernandez, E. Prefasi, Analog to digital conversion using noise shaping and time encoding. IEEE Trans. Circuits Syst. I **55**(7), 2026–2037 (2008)

32. L.R. Carley, J. Kenny, A 16-bit 4th order noise shaping D/A converter, in *IEEE CICC Proceedings*, Rochester, NY (1998), pp. 21.7.1–21.7.4

33. M. Kappes, H. Jensen, T. Gloerstad, A versatile 1.75 mW CMOS continuous-time delta-sigma ADC with 75 dB dynamic range for wireless applications, in *Proceedings of the 28th ESSCIRC* (2002)

34. M. Miller, C. Petrie, A multibit sigma-delta ADC for multimode receivers. IEEE J. Solid-State Circuits **38**(03), 475–482 (2003)

35. M. Ortmanns, F. Gerfers, *Continuous-Time Sigma-Delta A/D Conversion*. Springer Series in Advanced Microelectronics (Springer, Berlin, 2006)

36. M. Straayer, M. Perrott, A 12-bit, 10-MHz bandwidth, continuous-time $\Sigma\Delta$ ADC with a 5-bit, 950-MS/s VCO-based quantizer. IEEE J. Solid-State Circuits **43**(4), 805–814 (2008)

37. M. Bolatkale, L.J. Breems, R. Rutten, K.A.A. Makinwa, A 4 GHz continuous-time $\Delta\Sigma$ ADC with 70 db DR and 74 dBFS THD in 125 MHz BW. IEEE J. Solid-State Circuits **46**(12), 2857–2868 (2011)

38. N. DaDalt, M. Harteneck, Ch. Sandner, A. Wiesbauer, On the jitter requirements of the sampling clock for analog-to-digital converters. IEEE Trans. Circuits Syst. I **49**(9), 1354–1360 (2002)
39. N. Yaghini, D. Johns, A 43 mW CT complex $\Delta\Sigma$ ADC with 23 MHz of signal bandwidth and 68.8 dB SNDR, in *IEEE ISSCC Proceedings* (2005), pp. 502–503
40. O. Oliaei, P. Clement, Ph. Gorisse, A 5 mW $\Sigma\Delta$-modulator with 84 dB dynamic range for GSM/EDGE, in *IEEE ISSCC Proceedings* (2001)
41. P. Malla, H. Lakdawala, K. Kornegay, K. Soumyanath, A 28 mW spectrum-sensing reconfigurable 20 MHz 72 dB-SNR 70 dB-SNDR DT $\Delta\Sigma$ ADC for 802.11n/WiMAX receivers, in *IEEE ISSCC Proceedings* (2008), pp. 496–497
42. P. Pessl, R. Gaggl, J. Hohl, D. Giotta, J. Hauptmann, A four-channel ADSL2+ analog front-end for CO applications with 75 mW per channel, built in 0.13 μm CMOS. IEEE J. Solid-State Circuits **39**(12), 2371–2378 (2004)
43. P. Shettigar, S. Pavan, A 15 mW 3.6 GS/s CT-$\Delta\Sigma$ ADC with 36 MHz bandwidth and 83 dB DR 90 nm CMOS, in *IEEE ISSCC Proceedings* (2012), pp. 156–157
44. P. Witte, J.G. Kauffman, J. Becker, Y. Manoli, M. Ortmanns, A 72 dB-DR $\Delta\Sigma$ CT modulator using digitally estimated auxiliary DAC linearization achieving 88 fJ/conv in a 25 MHz BW, in *IEEE ISSCC Proceedings* (2012), pp. 154–155
45. Ph. Allen, D. Holberg, *CMOS Analog Circuit Design* (Oxford University Press, Oxford, 2002). ISBN 0-19-511644-5
46. R. Baird, T. Fiez, Improved $\Delta\Sigma$ DAC linearity using data weighted averaging, in *IEEE ISCAS Proceedings*, vol. 1 (1995), pp. 13–16
47. R. Gaggl, A. Wiesbauer, G. Fritz, Ch. Schranz, P. Pessl, A 85-dB dynamic range multibit delta-sigma ADC for ADSL-CO applications in 0.18 μm CMOS. IEEE J. Solid-State Circuits **38**(7), 1105–1114 (2003)
48. R. Gaggl, M. Inversi, A. Wiesbauer, A power optimized 14-bit SC $\Delta\Sigma$ modulator for ADSL CO applications, in *IEEE ISSCC Proceedings* (2004), pp. 82–83
49. R. Joerg, *Wireless LANs* (Heise Zeitschriften, Hannover, 2006)
50. R. Schoofs, *Design of High-Speed Continuous-Time Delta-Sigma A/D Converters for Broadband Communication* (Katholieke Universiteit Leuven, Belgium, 2007). ISBN 978-90-5682-845-5
51. R. Schoofs, M. Steyaert, W. Sansen, A 1 GHz continuous-time sigma-delta A/D converter in 90 nm standard CMOS, in *IEEE MTT-S Tech. Dig.* (2005), pp. 1287–1290
52. R. Schoofs, M. Steyaert, W. Sansen, A design-optimized continuous-time delta-sigma ADC for WLAN applications. IEEE Trans. Circuits Syst. I, Regul. Pap. **54**(1), 209–217 (2007)
53. R. Schreier, Delta-sigma toolbox for MATLAB. Technical report. http://www.mathworks.com/support/ftp/
54. R. Schreier, G. Temes, *Understanding Delta-Sigma Data Converters* (Wiley, Hoboken, 2005)
55. S. Kulchycki, R. Trofin, K. Vleugels, B. Wooley, A 77-dB dynamic range, 7.5-MHz hybrid continuous-time/discrete-time cascaded $\Sigma\Delta$ modulator. IEEE J. Solid-State Circuits **43**(4), 796–804 (2008)
56. S. Norsworthy, R. Schreier, G. Temes, *Delta-Sigma Data Converters: Theory, Design and Simulation* (Wiley-IEEE Press, New York, 1996). ISBN 0-7803-1045-4
57. S. Ouzounov, E. Roza, H. Hegt, G. van der Weide, A. van Roermund, An 8 MHz, 72 dB SFDR asynchronous sigma-delta modulator with 1.5 mW power dissipation, in *Proceedings 2004 Symp. VLSI Circuits* (2004), pp. 88–91
58. S. Ouzounov, H. Hegt, A. van Roermund, Sigma-delta modulators operating at a limit cycle. IEEE Trans. Circuits Syst. II **53**(5), 399–403 (2006)
59. S. Ouzounov, R. Veldhoven, C. Bastiaansen, K. Vongehr, R. Wegberg, G. Geelen, L. Breems, A. Roermund, A 1.2 V 121-mode CT $\Delta\Sigma$ modulator for wireless receivers 90 nm CMOS, in *IEEE ISSCC Proceedings* (2007), p. 242
60. S. Paton, Contribución al modelado y diseño de modulatores sigma-delta en tiempo continuo de baja relación de sobremuestreo y bajo consumo de potencia (English Summary). Technical report, Universidad Carlos III de Madrid, 2004; PhD thesis, Madrid, Spain

61. S. Paton, A. Giandomenico, L. Hernandez, A. Wiesbauer, Th. Poetscher, M. Clara, A 70-mW 300-MHz CMOS continuous-time sigma-delta ADC with 15-MHz bandwidth and 11 bits of resolution. IEEE J. Solid-State Circuits **39**(7), 1056–1063 (2004)
62. S. Paton, L. Hernandez, Low-voltage continuous-time sigma-delta modulator/research. Technical report, Universidad Carlos III de Madrid, Spain, 2002; internal report of consultation agreement 8500139 between Infineon and Universidad Carlos III de Madrid
63. St. Henzler, S. Koeppe, D. Lorenz, W. Kamp, R. Kuenemund, D. Schmitt-Landsiedel, A local passive time interpolation concept for variation-tolerant high-resolution time-to-digital conversion. IEEE J. Solid-State Circuits **43**(7), 1666–1676 (2008)
64. T. Brooks, D.H. Robertson, D.F. Kelly, A. Del Muro, S.W. Harston, A cascaded sigma-delta pipeline A/D converter with 1.25 MHz signal bandwidth and 89 db SNR. IEEE J. Solid-State Circuits **32**, 1896–1906 (1997)
65. T. Shui, R. Schreier, F. Hudson, Mismatch shaping for a current-mode multibit delta-sigma DAC. IEEE J. Solid-State Circuits **34**, 331–338 (1999)
66. Th. Burger, Q. Huang, A 13.5 mW, 185 M sample/s $\Delta\Sigma$-modulator for UMTS/GSM dual-standard IF reception, in *IEEE ISSCC Proceedings* (2001)
67. Th. Christen, Q. Huang, A 0.13 μm CMOS 0.1–20 MHz bandwidth 86–70 db DR multi-mode DT $\Delta\Sigma$ ADC for IMT-advanced, in *Proceedings of the 36th ESSCIRC* (2010), pp. 414–417
68. Th. Christen, Th. Burger, Q. Huang, A 0.13 μm CMOS EDGE/UMTS/WLAN tri-mode $\Delta\Sigma$ ADC with −92 dB THD, in *IEEE ISSCC Proceedings* (2007), pp. 240–241
69. V. Srinivasan, V. Wang, P. Satarzadeh, B. Haroun, M. Corsi, A 20 mW 61 dB SNDR (60 MHz BW) 1b 3rd-order continuous-time delta-sigma modulator clocked at 6 GHz in 45 nm CMOS, in *IEEE ISSCC Proceedings* (2012), pp. 158–159
70. V. Dhanasekaran, M. Gambhir, M.M. Elsayed, E. Sánchez-Sinencio, J. Silva-Martinez, C. Mishra, L. Chen, E.J. Pankratz, A continuous time multi-bit $\Delta\Sigma$ ADC using time domain quantizer and feedback element. IEEE J. Solid-State Circuits **46**(3), 639–650 (2011)
71. W. Sansen, *Analog Design Essentials* (Springer, Dordrecht, 2006)
72. W. Sansen, Laker, *Design of Analog Integrated Circuits ans Systems* (McGraw-Hill, New York, 1994)
73. W.Y. Chen, *DSL Simulation Techniques and Standards/Development for Digital Subscriber/Line Systems* (Macmillian Technical, Indianapolis, 1998)
74. W. Yang, W. Schofield, H. Shibata, S. Korrapati, A. Shaikh, N. Abaskharoun, D. Ribner, A 100 mW 10 MHz-BW CT $\Delta\Sigma$ modulator with 87 dB DR and 91 dBc IMD, in *IEEE ISSCC Proceedings* (2008), pp. 498–499
75. Y. Geerts, Design of high-performance CMOS delta-sigma A/D converters. Technical report, ETAS-MICAS, K.U. Leuven, 2001; PhD thesis, Belgium, December 2001
76. Y. Geerts, A. Marques, M. Steyaert, W. Sansen, A 3.3 V 15-bit delta-sigma ADC with a signal bandwidth of 1.1 MHz for DSL-applications. IEEE J. Solid-State Circuits **34**(7), 927–936 (1999)
77. Y. Geerts, M. Steyaert, W. Sansen, A high-performance multi-bit CMOS delta-sigma converter. IEEE J. Solid-State Circuits **35**(12), 1829–1840 (2000)
78. Y. Shu, B. Song, K. Bacrania, A 65 nm CMOS $\Delta\Sigma$ modulator with 81 dB DR and 8 MHz BW auto-tuned by pulse injection, in *IEEE ISSCC Proceedings* (2008), pp. 500–501

Index

R. Gaggl, *Delta-Sigma A/D-Converters*, Springer Series in Advanced Microelectronics 39,   143
DOI 10.1007/978-3-642-34543-2, © Springer-Verlag Berlin Heidelberg 2013

MIX
Papier aus verantwortungsvollen Quellen
Paper from responsible sources
FSC® C105338

FSC
www.fsc.org

If you have any concerns about our products,
you can contact us on
ProductSafety@springernature.com

In case Publisher is established outside the EU,
the EU authorized representative is:
Springer Nature Customer Service Center GmbH
Europaplatz 3, 69115 Heidelberg, Germany

Printed by Libri Plureos GmbH
in Hamburg, Germany